Ionic Polymer–Metal Composites

Ionic Polymer–Metal Composites
Evolution, Applications and Future Directions

Edited by
Srijan Bhattacharya

CRC Press is an imprint of the
Taylor & Francis Group, an **informa** business

First edition published 2022
by CRC Press
6000 Broken Sound Parkway NW, Suite 300, Boca Raton, FL 33487-2742

and by CRC Press
2 Park Square, Milton Park, Abingdon, Oxon, OX14 4RN

© 2022 selection and editorial matter, Srijan Bhattacharya; individual chapters, the contributors

CRC Press is an imprint of Taylor & Francis Group, LLC

Reasonable efforts have been made to publish reliable data and information, but the author and publisher cannot assume responsibility for the validity of all materials or the consequences of their use. The authors and publishers have attempted to trace the copyright holders of all material reproduced in this publication and apologize to copyright holders if permission to publish in this form has not been obtained. If any copyright material has not been acknowledged please write and let us know so we may rectify in any future reprint.

Except as permitted under U.S. Copyright Law, no part of this book may be reprinted, reproduced, transmitted, or utilized in any form by any electronic, mechanical, or other means, now known or hereafter invented, including photocopying, microfilming, and recording, or in any information storage or retrieval system, without written permission from the publishers.

For permission to photocopy or use material electronically from this work, access www.copyright.com or contact the Copyright Clearance Center, Inc. (CCC), 222 Rosewood Drive, Danvers, MA 01923, 978-750-8400. For works that are not available on CCC please contact mpkbookspermissions@tandf.co.uk

Trademark notice: Product or corporate names may be trademarks or registered trademarks and are used only for identification and explanation without intent to infringe.

Library of Congress Cataloging-in-Publication Data
A catalog record has been requested for this book

ISBN: 978-1-032-06945-6 (hbk)
ISBN: 978-1-032-06946-3 (pbk)
ISBN: 978-1-003-20466-4 (ebk)

DOI: 10.1201/9781003204664

Typeset in Times
by codeMantra

Contents

Preface vii
Acknowledgments ix
Biography xi
Contributors xiii

Chapter 1 Introduction to IPMC, Its Application and Present Scenario 1

Srijan Bhattacharya

Chapter 2 Ionic Polymer–Metal Composite Actuators:
Methods of Preparation 17

Dillip Kumar Biswal

Chapter 3 A Study of Movement, Structural Stability, and
Electrical Performance of a Harvesting System Based
on Ionic Polymer–Metal Composites 31

Nang Xuan Ho, Vinh Nguyen Duy, and Hyung-Man Kim

Chapter 4 Application of Ionic Polymer Metal Composite (IPMC) as
Soft Actuators in Robotics and Bio-Mimetics 53

Ravi Kant Jain

Chapter 5 Inverse Kinematic Modeling of Bending Response of Ionic
Polymer Metal Composite Actuators 95

*Siladitya Khan, Gautam Gare, Ritwik Chattaraj,
Srijan Bhattacharya, Bikash Bepari, and Subhasis Bhaumik*

Chapter 6 Selection of Elastomer for Compliant Robotic Gripper
Harnessed with IPMC Actuator 123

Srijan Bhattacharya, Bikash Bepari, and Subhasis Bhaumik

Chapter 7 Study of Polar Region Atmospheric Electric Field Impact
on Human Beings and the Potential Solution by IPMC 149

*Suman Das, Srijan Bhattacharya, and
Subrata Chattopadhyay*

Chapter 8 Future Directions on IPMC Research 193
Srijan Bhattacharya

Index 199

Preface

It is well known that modern robotics and numerous other applications need to have artificial muscles. Such applications also demand the need of soft actuators. Why is this so? The simple reason for this is that this is how the human body works. Human body is probably one example of extreme engineering governed by the Mother Nature through a very complicated control that embraces biophysics, biomechanics, biochemistry, and self-actuation. While it is yet to be possible to exactly mimic it, the human endeavor in search of suitable materials continues to grow. So far, many materials have been tried upon. These include, but are not necessarily limited to, liquid crystal elastomers and dielectric elastomers. These attempts also include bilayer hydrogels. Further attempts are also made using the bilayer nanomaterials. Furthermore, efforts are on to utilize highly oriented polymer fibers. The simple principle that governs their choices is that they can respond to various external stimuli. For instance, the stimulant could be heat. Similarly, it could also be the presence of light as well as the electric field. Even the presence of water can act as a stimulant.

Therefore, it is not surprising at all to note that a huge amount of research is globally directed toward the development of various soft actuators appropriate for a wide variety of applications. Thus, the ionic polymer–metal composites (IPMCs) are a class of new soft actuator materials being developed globally to address the issues related to soft actuation. These have a layered and/or sandwich structure. In fact, it is generally true that in IPMCs, an ion exchange membrane is sandwiched between two appropriate electrodes. They are very promising new materials as they can provide decent magnitude of bending-induced actuation. This can really happen even at a voltage as small as, e.g., 5 V. This process is believed to happen due to fast ion migration. This is what gives the IPMCs a unique advantage because their actuations can be precisely controlled by the application of an external electric field.

The book offers a well-planned documentation of the latest research in IPMCs. It is unique in nature and stands out in terms of the rich contents that it has to offer not only for the advanced researchers but also for the beginners. Thus, it spans a range that covers the classic academia on the one hand and the industries on the other. There is no doubt in my mind that it will also serve to generate a lot of interest not only in researchers of today but also in researchers of tomorrow.

The first chapter deals with an interesting overview of a very special class of composite materials known as ionic polymer–metal composites (IPMCs). The polymeric component in most of the case studies is an electroactive polymer (EAP). Thus, the overview summarizes many important applications of the IPMCs, such as the biomedical engineering, materials science, robotics, other industrial applications, and underwater applications.

The second chapter deals with mainly the methods, e.g., the chemical decomposition, to prepare the IPMCs and their bending characterization as a function of the applied voltage. It establishes that mostly Nafion-117 and rarely Flemion are employed as the foundation polymer matrix and conducting materials such as platinum, gold, and silver are used as the electrode materials to fabricate the IPMCs.

The third chapter presents a simulated IPMC-based electrochemical energy harvesting system installed in the ocean and produced using the computational fluid dynamics (CFD) method. The simulation processes focus on the movement and structural stability of the system design in the sea to protect the IPMC module against possible damage, which would directly affect the power output. Furthermore, the experimental tests under natural marine conditions are also utilized to analyze the electrical harvesting performance of the IPMC system. These results show that the use of IPMC materials has many advantages as they are soft and durable. Thus, it is emphasized that they can respond faster to wave parameters such as frequency, amplitude, and wavelength.

The fourth chapter focuses on the development, fabrication method, and characterization of SWNT-based IPMC soft actuators. It is shown that this unique class of IPMCs provides many advantages such as high proton conductivity, high ion exchange, high water uptake, high tensile strength, good thermal properties, excellent film forming capacity, large bending, and electromechanical and bending properties.

On the other hand, the fifth chapter summarizes an alternate overview of simulation methodologies that have emerged to explain the sensing and actuation attributes of smart materials with a specific emphasis on mechanics-based simulations. It is shown that such methodologies are capable of successfully explaining high-DoF (degree of freedom), underactuated attributes in addition to high deformability upon meager activations that are exhibited by such materials.

The sixth chapter reviews the issues related to the selection of elastomer for a compliant robotic gripper harnessed with IPMC actuators.

The seventh chapter gives a new idea of the use of IPMCs in Polar Regions. Finally, the eighth chapter gives a broad overview of the future research domains of IPMCs. It emphasizes that the emerging prospects of IPMCs are very exciting, and it should be possible to address many of these issues in days to come such that a plethora of new application domains can roll out successfully. I wish this great effort all the very best.

Dr Anoop Kumar Mukhopadhyay
Professor, Department of Physics, School of Basic Sciences
Dean, Faculty of Science, Manipal University Jaipur
Jaipur 303007, Rajasthan, India
Jaipur, the 18th of August 2021

Acknowledgments

It is my privilege to express my profound and sincere thanks to all the authors from different countries (India, the USA, Vietnam, and South Korea) who have contributed to this book—"Ionic Polymer–Metal Composites: Evolution, Applications and Future Directions". My special thanks to Dr. Arjun Dey, Scientist, ISRO, for his continuous support and encouragement to complete this project; without his support, it would have been quite impossible for me to complete this project.

Thanks to my respected doctoral thesis supervisors and research guides Dr. Subhasis Bhaumik, Dean (Research & Consultancy), Professor, Aerospace Engineering & Applied Mechanics Department (AE & AM), Indian Institute of Engineering Science & Technology (IIEST), Shibpur, and Dr. Bikash Bepari, Professor, Production Engineering Department, Haldia Institute of Technology, Haldia, for their continuous support and encouragement.

I also offer my sincere thanks to the management of RCC Institute of Information Technology, Kolkata, India, for giving me the opportunity to complete this project, and also to my Applied Electronics & Instrumentation Departmental colleagues for their continuous support and encouragement.

Lastly, I would like to acknowledge my debt to my parents Dr. Sanat Kumar Bhattacharya and Dr. Tapasi Bhattacharya and my beloved wife Smt. Piyali Bhattacharya for their tireless encouragement and support to complete this book in due time.

Dr. Srijan Bhattacharya
RCC Institute of Information Technology,
Kolkata, India

Biography

Dr. Srijan Bhattacharya (M'16, SM'20) is working as Assistant Professor and Head in Department of Applied Electronics & Instrumentation Engineering (AEIE) in RCC Institute of Information Technology, Kolkata, India. He did B.E. in Electronics & Instrumentation Engineering in Gandhi Institute of Engineering & Technology (presently GIET University), Orissa, India (2003), M.Tech. in Electrical Engineering (Specialization in Mechatronics) in National Institute of Technical Teachers Training & Research, Kolkata, India (2008), and PhD in Indian Institute of Engineering Science and Technology, Shibpur, India (2017). He received a gold medal from West Bengal University of Technology, West Bengal [presently Maulana Abul Kalam Azad University of Technology (MAKAUT)]. He had published a book titled *Advancements in Instrumentation and Control in Applied System Applications* as a single editor with IGI Global, USA, and a conference proceedings with the book series Lecture Notes in Electrical Engineering, Springer.

He is an Associate Guest Editor of the Springer Journal of *The Institution of Engineers (India): Series B*. He has published 30+ peer-reviewed journals/book chapters/conference papers and completed one research project sponsored by Institution of Engineers (India)—IE(I). He has been involved in academic administration as Board of Studies Member, AEIE MAKAUT, West Bengal, GIET University, India. He is also a coordinator of the Student Chapter RCCIIT—The Robotics Society (TRS)—India. He is also a Life Member of the Institution of Engineers (India) and The Robotics Society (TRS), India. His research interest is in mechatronics system design, sensors design and application, and smart materials—ionic polymer–metal composites (IPMCs).

Contributors

Dr. Bikash Bepari
Department of Production Engineering
Haldia Institute of Technology
Haldia, India

Dr. Srijan Bhattacharya
Applied Electronics and Instrumentation Engineering
RCC Institute of Information Technology
Kolkata, India

Dr. Subhasis Bhaumik
Aerospace Engineering and Applied Mechanics
Indian Institute of Engineering Science and Technology
Howrah, India

Dr. Dillip Kumar Biswal
Department of Mechanical Engineering
C V Raman Global University
Bhubaneswar, India

Dr. Ritwik Chattaraj
Prestige Tech Park
Bengaluru, India

Dr. Subrata Chattopadhyay
Electrical Engineering
NITTTR
Kolkata, India

Mr. Suman Das
Electrical Engineering Department
MCKV Institute of Engineering
Howrah, India

Dr. Vinh Nguyen Duy
Faculty of Vehicle and Energy Engineering
Phenikaa University
Hanoi, Vietnam

Gautam Rajendrakumar Gare
Carnegie Mellon University
Pittsburgh, USA

Dr. Nang Xuan Ho
Faculty of Vehicle and Energy Engineering
Phenikaa University
Hanoi, Viet Nam

Dr. Ravi Kant Jain
Information Technology Group, Engineering Services Division (ESD-Institute)
CSIR-Central Mechanical Engineering Research Institute (CMERI)
Durgapur, India

Siladitya Khan
Department of Biomedical Engineering
University of Rochester
Rochester, USA

Dr. Hyung-Man Kim
Power System and Sustainable Energy Laboratory
Inje University
Gimhae, South Korea

1 Introduction to IPMC, Its Application and Present Scenario

Srijan Bhattacharya
RCC Institute of Information Technology
Kolkata, India

CONTENTS

1.1 Introduction 1
1.2 Literature Survey 4
    1.2.1 Electrical Characterization of IPMC 5
    1.2.2 Control Issues of IPMC 7
    1.2.3 IPMC Grippers 8
    1.2.4 Compliant Materials for Microgripper 9
    1.2.5 IPMC Applications in Space 10
1.3 Summary 11
References 11

1.1 INTRODUCTION

When handling small objects of which one of the dimensions is less than 1 mm, it becomes a challenge for the engineers to develop a suitable gripping device that can serve the purpose efficiently. This results in the study of micro-object manipulation and the design of microgrippers or gripping devices, which is the most challenging task for the scientific community throughout the globe. At the same time, micro-object handling has been solved by many of the researchers in the past few decades. In the past literature, it has been observed that in the field of microgripper design most of the research work was done using two-jaw gripping devices, using materials such as metals (spring steel, aluminum, etc.) or silicon rubber (usually polydimethylsiloxane). The actuation for those gripping devices is controlled externally. For this, a technique is adopted for handling lightweight objects. A human finger-like structure is designed using polydimethylsiloxane (PDMS) as the base material, and the ionic polymer–metal composite (IPMC) is used as the smart actuator for actuating the finger. Characterization of various shapes and sizes of IPMCs is done before the operation for finger actuation. The behavior of IPMCs after soaking in various ionic solvents is also experimented. These solvents are used before testing the IPMC as an actuator. To get better

DOI: 10.1201/9781003204664-1

force and displacement as output from the IPMC, statistical analyses are also performed on the experimental results of the IPMC. Tele-operation is reported by the researchers for a two-jaw IPMC gripper, where a micronewton force calculation technique has been used with an imaging technique.

In addition to compliance, less weight-to-volume ratio and ease of fabrication, electroactive polymers (EAPs) exhibit a significant change in physical appearance when subjected to electrical stimuli, which motivates the scientific community throughout the globe to apprehend the potential benefit to be used as an actuator [1]. The prima facie attribute of EAPs is their ability to act as a substitute to the conventional actuators and sensors, consuming less energy [2]. EAPs are generally classified into two types: (1) electronic and (2) ionic EAPs, depending upon their activation mechanism. Electronic EAPs require high activation voltage and eventually exhibit much more mechanical actuation energy. On the contrary, ionic EAPs are activated by diffusion of ions as they always consist of two electrodes [3]. Compared to electronic counterparts, ionic EAPs exhibit larger electromechanical strain when subjected to an electrical stimulus, and this is similar to the behavior of biological muscles in terms of articulating force and stress. For this reason, EAPs are termed as artificial muscles [4].

As discussed earlier, IPMCs, a kind of ionic EAPs (usually Nafion® and Flemion®), are a polyelectrolyte ion exchange membrane plated on both the faces with a noble electrode material, and it is countered by certain change ions. The actuation of an IPMC beam, when subjected to voltage across the thickness, is primarily dominated by ion and water content redistribution and drifting of cations toward the cathode, while the anions remain stationary. Subsequently, the cations and water molecules assemble near the cathode, whereas a paucity of molecule concentration is found near the anode, yielding both concentrated and weak boundary layers [6]. Figure 1.1 illustrates the cross section of an IPMC beam and the actuation mechanism.

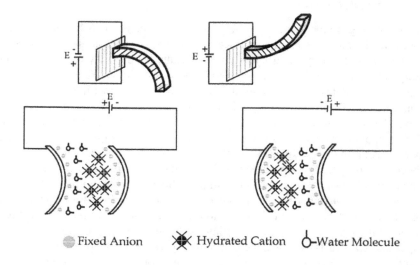

FIGURE 1.1 IPMC actuation mechanism.

Introduction to IPMCs and Applications

The upshots (responses) of IPMC actuation are mainly the actuation force, displacement, response speed and endurance from the user point of view, which play an essential role in actuator design [1]. Out of these desired outputs, the actuation force and displacement are more worthwhile from the application point of view [7].

It has been reported that an increase in the applied voltage increases both the bending displacement and the force. The linearity of bending displacement and force depends on the thickness of IPMC strips. As the thickness reduces, it creates more linearity, but contrastingly generates less force. In addition, the span length and thickness have immense effects on the actuation force and displacement. The actuating force is linearly proportional to the width and the square of the thickness and inversely proportional to the length, as reported by [8]. Furthermore, the elastic modulus of the IPMC is proportional to the thickness as well as applied frequency, as reported by [1] and [9], respectively. Eventually, high thickness and frequency both increase the elastic modulus; hence, the displacement is less. Contrastingly, as claimed by [9], a higher thickness gives rise to a high actuation force.

Ref. [10] reported an exhaustive characterization of mechanical properties for span length and width of IPMC strips in actuation of anthropomorphic microgripper. In the recent past, micro-object manipulation and design of microgripper have gained paramount importance as revealed in the literature survey. In micromanipulation, the size of the manipulated object is usually much less than 1 mm in at least one dimension, as revealed by [11]. The manipulation of biological cells was addressed with three miniature grippers with parallel and angular jaw motions, which can also measure forces. Of them, one was made with implicit manufacturing constraints through topology optimization. In their work, the gripper was used to hold a yeast ball of less than 1 mm in diameter and the forces involved therein were estimated to be about 30 mN. Micromanipulation eventually requires an actuator that can produce a minuscule force which shall not deform the stature of the object during manipulation. The different forces that come into action during microgripping, such as van der Waals forces, electrostatic forces and surface tension forces, were also discussed. They proposed to use the laser Raman spectrophotometer for the measurement and calibration of the force applied at the micro-object. Eventually, many issues and aspects are evolving into the scenario of IPMC-assisted micromanipulation and are revealed gradually in the next section. IPMCs being one of the popular EAPs have been considered to act either individually or in conjugation with a PDMS-based compliant gripper. PDMS as a compliant gripper material comes to the forefront as it is biocompatible and substantive with a methodological approach. Characterization of IPMC is another prima facie concern that corroborates both the mechanical and electrical aspects. The mechanical aspects include tractrix (locating the tip position with certainty), equivalent beam model, variation in physical attributes (length, breadth and width) and obviously the anticipation of the behavior in doping solutions. The electrical counterpart of characterization includes the response of IPMC when it is subjected to different frequency and voltage levels. The control issue entails the mathematical modeling so as to fetch the upshots by tweaking the different gains pertinent to PID control. In addition, an attempt has already been reported for joystick control keeping in view of tele-operation, also in biomedical application. Paul et al. [12] and Bhattacharya et al. [13] reported that the IPMC can be used as an EMG sensor and it can also record the information of different human arm activities.

1.2 LITERATURE SURVEY

As the present investigation revolves around many aspects of IPMC actuation, the following literature survey reveals the advantages and disadvantages of research pertinent to the said field.

The authors of the articles [4,5] showed the use of EAPs as artificial muscles. A comparison between EAPs, EACs (electroactive ceramics) and SMAs (shape-memory alloys) has been made, and it has been found that the properties of EAPs offer superior capability.

The EAPs can be used to develop artificial muscles with the potential of developing biologically inspired robots that can possibly walk, fly, hop, dig, swim and/or drive. Different types of EAPs are wet (ionic) EAP ionic polymers, IPMCs, carbon nanotubes, dry EAP ferroelectric polymers (PVDF or PVF2), electrostatically stricted polymer (ESSP) actuators and electro-viscoelastic elastomers. They also studied the characteristics of IPMC of dimensions 7.7 mm×3.3 mm×0.196 mm, and they studied the response under water at 6.5 mm from the sample tip, where a resonance is observed at 113 Hz in water and at 272 Hz in air. They also developed image processing algorithms such as (1) clipping, (2) edge detection and (3) surface tracing of the IPMC strips. The functional potential of IPMC sensors for measuring bending angles and rates in the context of hand prosthetics was also studied. EAP sensor types that have been described by the authors are polypyrrole (PPy), PVDF and IPMC. Brunetto et al. [14] showed that IPMCs can be used as actuators and sensors. IPMCs work at low actuation voltage from 0 to 5 V at low frequency from 1 to 20 Hz. IPMCs are resistive (>50 Ω) in the high-frequency range. And they are capacitive (>100 μF) in the low-frequency range. (It has been reported in the literature that the equivalent circuit of an IPMC is a series combination of resister and capacitor, so the equivalent impedance $Z = R + \dfrac{1}{j2\pi fC}$. At high frequencies, f tends to ∞ and $Z=R$, so we can say that it is resistive at high frequencies. But at low frequencies, the capacitance value becomes infinity, which shows the capacitance nature of IPMCs at low frequencies.) The load-carrying capacity is found to be inversely proportional to the length of the IPMC.

He et al. [6] reported that the blocking force increases with the increase in the IPMC membrane thickness, the displacement decreases with the increase in the IPMC membrane thickness, the current increases with the increase in the IPMC membrane thickness, and the elastic modulus increases with the increase in the IPMC membrane thickness. To achieve blocking force, the displacement is more

Introduction to IPMCs and Applications

for the less thick IPMC membrane. Leary et al. [15] used an IPMC sample of dimensions 30 mm×5 mm×0.2 mm and specifications similar to Nafion #117 (DuPont) with platinum electrodes and mobile Na⁺ counterions to explain the electrical impedance of IPMCs. Applying a cosine wave of 7 V and 0.03 Hz to the sample exhibits a nonlinear hysteresis behavior. It has been concluded that a large bending actuation occurs at low voltage level (4 V and steady-state current 0.6 A).

Kim et al. [17] measured the stiffness, displacement and force of an IPMC (dimensions 30 mm×4 mm×0.4/1.2 mm). The input given to the IPMCs is a 2–5 Vpp AC sine wave with 0.5 Hz frequency. The stiffness of the IPMC is measured with the governing equation $\delta = \dfrac{PL^3}{3EI}$, where δ is the measured tip displacement of the actuator, P is the force applied to the actuator, L is the length of the actuator, E is Young's modulus, and I is the moment of inertia. The stiffness of the sample is proportional to the third power of the thickness $EI = E\dfrac{bh^3}{12}$, where b is the width and h is the thickness of the IPMC actuator. It is seen from the above equation that the stiffness and force increase with the increasing thickness of the IPMC, but the displacement increases with decreasing thickness. Wu et al. [18] of University of California verified the proposed micromechanical model with Nemat-Nasser's model by applying 1 V to IPMC increase time 0 s to 10 s to 20 s. The total transported charge by cation during actuation was obtained by the time integration of the measured current, which gradually increases to 9.2 mA at 10.0 s and then decreases to 0.26 mA. Lee et al. [19] fabricated an IPMC with a greater thickness. After fabricating the IPMC, it was plated with platinum, and if the surface resistance of the IPMC decreases, the tip deflection increases with the application of driving voltage. Before and after, the thickness of the IPMC was taken as 0.51 –0.58 mm, 0.67 –0.78 mm and 0.84 –0.97 mm. The tip force of the IPMC increases with the increase in the applied voltage when the thickness of the IPMC increases. The Young's modulus of the IPMC increases with the thickness of the IPMC because the tensile stress (force/unit area) increases compared to the tensile strain. Lee showed that the IPMC can be used as artificial fingers.

1.2.1 Electrical Characterization of IPMC

Yu et al. [20] fabricated an IPMC that is twice plated with platinum. After plating, Pt penetrates through the membrane of 9 μm thickness. The specimen IPMC used here had the dimensions of 30 mm×3 mm×0.2 mm. A charge-coupled device (CCD) camera was used for the displacement measurement. LabVIEW was used for controlling the power supply as square wave, sinusoidal wave and triangular wave, with supply frequency ranging from 0.1 to 100 Hz and voltage less than 10 V. The IPMC strip bends toward the anode side when an electrical field is applied. From the experimental results, it can be concluded that with the same applied voltage and frequency, the square wave gives the maximum displacement.

With the increase in frequency, force and displacement decrease. The performance of the IPMC is reasonable and increases in the order square, sinusoidal and triangular waves, respectively. Stoimenov et al. [21] developed an IPMC with different electrode areas in the membrane surface, and they observed different curvatures of the IPMC applying a potential of 2 V. They also experimented on a bistable buckled IPMC beam, S-shaped curve with varying curvature of the end segments in IPMC strip. They also proposed some applications such as forming of gel-like substance into complex shapes by IPMC nozzle and heart compression device for applying local pressure. Porfiri [22] explained the charge dynamics in IPMCs. The circuit conductivity of the IPMC is proportional to the ion diffusivity, where the dielectric constant is independent. The capacitance of the IPMC is proportional to the square root of the dielectric constant, where the ion diffusivity is independent. With all these results, Porfiri [22] developed the equivalent circuit (series RC) of an IPMC of finite surface area. The capacitance is independent of IPMC thickness, and the thickness linearly affects the equivalent resistance. These electrostatic interactions lead to the bending moment of the IPMC. At low voltage levels (100 mV), the capacitance of the IPMC becomes prominent, and as voltage increases, the anode–polymer interface growth for this IPMC capacitance decreases drastically in the order of 1/5 of its initial value. It is explained that the cations are free to move in the polymer region, whereas the anions are fixed with the backbone polymer. Brunetto et al. [16] showed that the characteristics of IPMC change with the change in temperature and humidity as Young's modulus of the IPMC also changes. They developed a model of IPMCs with applied voltage and current. One portion of the current produces charge redistribution and a stress inside the IPMC. This stress can be divided into two components: a blocked force and a free deflection. Chen et al. [23] provided the mathematical modeling of IPMC behavior as nonlinear capacitance. They made this analysis with a partial differential equation (PDE) and showed that the DC polymer resistance and pseudo-capacitance are not included by Profiri. In their model, Ca is the pseudo-capacitance, metal ion (Na^+ or Ca^+) diffusion resistance, R_{c2} is the hydrogen ion diffusion resistance, R_a is the electrode resistance, and R_{dc} is the nonlinear DC resistance of the polymer. Paquette et al. [24] described the behavior of IPMCs in multilayer configuration. Three identical squares of IPMC strips of dimensions 1 cm × 1 cm × 0.2 cm are sandwiched for use in this application. Here, it is seen that the response time is much lower than that of a typical IPMC. In this electrochemical process, the impedance behavior of the IPMC and the equivalent circuit and schematic representation of the multilayered configuration of the IPMC were described with the circuit propagating in the X- and Z-directions.

Lopas et al. [25] described a model of IPMC where the IPMC is not uniformly plated; i.e., it is not uniformly charged. This model is compared with the finite element method (FEM). Among the two electrolytes, the IPMC with propylene carbonate plus lithium ion electrolyte had more displacement (5.10 cm) than the IPMC with water plus sodium ions as electrolyte (0.88 mm for the applied current and voltage of 1 mA and 1.5 V). They modeled the IPMC in two ways: mechanical and electrical models. Junga et al. [26] showed the performance of the IPMC

Introduction to IPMCs and Applications

actuator varying the frequency of the actuation voltage, where the size of actuator is 10 mm×2 mm×0.2 mm and a square input voltage of ±1.5Vp–p is applied. The IPMC actuator generates most of its deformation just after the polarity of the excitation voltage is switched in different applied signals such as square, sinusoidal and triangular waves. The square input consumes quite a large amount of power than the others. Driving the IPMC actuator by the sinusoidal input is more efficient than using the square one with respect to the power consumption.

1.2.2 Control Issues of IPMC

Jain et al. [27] experimented on a 40 mm×10 mm×0.2 mm IPMC strip with the EMG signal from the index finger. The EMG signal amplified 2550 times was sent to a PID controller, and it gave a deflection of 12 mm. When a potential of 3 V is applied to the IPMC, it gives 10 mN force and 12 mm displacement. Lee et al. [19] reported that the force increases and the bending decreases with the increase in the thickness of the IPMC. They controlled the IPMC-based artificial muscle using the electromyographic (EMG) signals obtained from the human forearm. The EMG signals generated by an intended contraction of muscles in the forearm were used for the actuation of the IPMC. Shan et al. [28] explained how tracking error can be reduced by applying feedforward dynamics compensation in IPMC actuators. The movement of the IPMC cannot track high-frequency electrical inputs efficiently due to dynamic effects, resulting in tracking error. A PI feedback controller in conjunction with a feedforward controller can be used to eliminate the unmeasured disturbances coming due to the unmodeled structure of the IPMC. To apply the feedforward control, the linear dynamic model is found experimentally by taking into consideration the frequency response. Brunetto et al. [15] developed a fractional-order model that is able to predict accurately the behavior of the IPMC actuator according to the variations in the length of the IPMC strip. Marquardt algorithm was developed in the MATLAB® environment. The IPMCs used were Nafion® #117 IPMCs and were 25, 30 and 35 mm long; 4 mm wide; and 200 mm thick. Xue et al. [29] described a model for the control of IPMC for micromanipulation. To control the slipping over the object, the tip deflection of the IPMC is designed with a linear–quadratic regulator (LQR) controller. The IPMC is used as an actuator, and the PVDF is used as a sensor. Both of them are of the same specifications and of dimensions 10 mm×5 mm×0.3 mm. Richardson et al. [30] gave an idea to control IPMC actuators. The IPMC output is measured in terms of either position or force. In this research, the authors dealt with the position of the IPMC. A PID position control was derived for open- and closed-loop control. The response time in the case of open-loop control was more than that in the case of the closed-loop system. Also a small overshoot was observed in the case of closed-loop system. The system was represented as a spring, mass and damper system. The actual position of the IPMC was measured with a laser sensor. Tsiakmakis et al. [31] recorded a motion parameter with a CCD video camera. First, the motion was captured in the video and then the captured video was transferred into frames and analyzed. The camera-based measurement provided more prominent results than the laser displacement measurement. With the application of 0–7 Hz frequency, the maximum displacement observed was 0–3.5 mm. This method is quite suitable for underwater actuation measurement. Bonomo et al. [32] showed an application of integrated motion of IPMC (30 mm by 5 mm) sensors and actuators. In this, one IPMC sensor and one actuator were placed in a plastic sheet stimulated externally and the voltage output was sensed, then amplified and executed for the closed-loop system. They also concluded that in the closed-loop system, the settling point was less than in the open-loop system. When positive feedback is given to the closed-loop system, it starts oscillating at its natural frequency.

1.2.3 IPMC Grippers

Shahinpoor et al. [33] presented a review of an IPMC application. The IPMC can be used as a microgripper, using a four-fingered gripper. They also gave an idea of the three-dimensional actuator concept. They also showed that the IPMC can be used in robotic fish design, highlighting the design of the fin for fish robots. They also noted that the IPMC can be used as a linear actuator or spiral actuator. A double-diaphragm pump of dimensions 5 mm × 1 mm × 0.2 mm was also developed using the IPMC. The IPMC artificial muscles can be used as human muscle power. A heart compression device can also designed with IPMC actuators. It is also shown that the IPMC can be used to activate bionic vision.

Feng et al. [34] provided a new idea of fabricating IPMCs and designed a μIPMC of dimensions 6 mm × 6 mm × 80 μm for biological applications. It gave the displacement and force outputs of 300 μm and 5 mN, respectively. The supply given to the system was a 12 V, 0.5 Hz square wave. This μIPMC gripper grips a flexible tube of 800 μm. Jain et al. [35] developed an IPMC-based two-finger microgripper which can grip a 10-gram object. From the experimental results, it is seen that within the supply voltage from 0 to 3.5 V, the IPMC of dimensions 30 mm × 6 mm × 0.2 mm generates a force of 0 to 0.11 mN. Jain et al. [36] developed a remote center compliance (RCC)-based microgripper with an IPMC actuator for peg-in-hole (PIH) microassembly operations. The IPMC microgripper has three fingers of dimensions 40 mm × 10 mm × 0.2 mm. It is seen that the insertion depth increases with the applied voltage. Lumia et al. [37] developed an IPMC actuator-based microgripper. This type of gripper has a load-carrying capability depending upon the force the fingers exert and the length-to-width ratio of the fingers. This type of microgripper structure is known as a pincher. This pincher is attached to a spring for opening and closing the gripper. The IPMC is used as a finger with a gold-plated electrode. The IPMC finger dimensions are 5 mm × 1 mm × 0.2 mm, which can carry a rigid micro-object made of solder of size 100–1500 μm in diameter. The mass of the microball is 15×10^{-6} kg, and the force exerted by the IPMC fingers can be 85 μN, which shows the load-carrying capability of the IPMC. From this, it can be seen that the IPMC can be used in different shapes and sizes. Through the experiment, it is shown that the IPMC can be used as a sensor in practical applications. Shahinpoor et al. [38] provided

Introduction to IPMCs and Applications

a review of ion exchange polymer–noble metal composite IPMCs as biomimetic sensors and actuators. An end effector gripper was developed for lifting a 10.3 g rock under 5 V, 25 mW activation using four 0.1 g fingers made of IPMCs. Their resistance increased with decreasing temperature, a property that is opposite to all metallic conductors. Chen et al. [39] developed a novel hybrid IPMC/PVDF (polyvinylidene fluoride) structure for both micro-sensing and micro-actuation. An open-loop injection experiment with the IPMC/PVDF sensori-actuator was conducted, and the process of the injection behavior was captured by the PVDF sensor. A micropipette with an ultra-sharp tip (1.685 µm in diameter and 2.65° in angle) was mounted at the end point of a rigid tip attached to the IPMC/PVDF structure. Sahoo et al. [40] described IPMCs as EAPs and their applications. Conducting polymer as polypyrrole (PPy) with gold (Au) i.e., PPy/Au layer-based devices were used as micro-actuators, which can be used as fingers. It is proposed to be used in cell manipulation and can be used in the field of chemistry. The applications of IPMC actuators are recorded for dust wiper and as gripper. A permanent deformation is observed in the IPMC when a DC voltage is applied to it. Under AC activation, the IPMC operates efficiently. Deole et al. [41] developed a microgripper with an IPMC of dimensions 5 mm × 1 mm × 0.2 mm and measured a load-carrying capacity of 15 mg. The weight of each IPMC finger was measured to be 2.1 mg. Thus, the IPMC fingers were able to lift a weight more than seven times the weight of each finger. The results show that the load-carrying capacity of a microgripper (using IPMC) varies with the finger length linearly. As the length of the finger decreases, the amount of tip displacement also decreases. Consequently, the microgripper with shorter fingers must be positioned more accurately with respect to an object than the microgripper with longer fingers.

1.2.4 Compliant Materials for Microgripper

Ananthasuresh et al. [42] designed a compliant microgripper to handle biological cells. This gripper is made of PDMS and spring steel. The microgripper was made from spring steel of size 11 mm × 8 mm × 0.5 mm using wire-cut electrical discharge machining (EDM). They tested this gripper with spherical-shaped zebra fish egg cells (nearly 0.7 mm in diameter), ellipsoidal-shaped drosophila (fruit fly) embryos (approx. 0.2 mm wide and 0.5 mm long), yeast balls (less than 1 mm in diameter) and hibiscus pollen (0.1 mm in diameter). Hegde et al. [43] showed the spring-leverage model for the compliant structures and developed a user interface using MATLAB with this they introduced a design methodology for compliant mechanisms for practical applications.

Rao et al. [44] developed a 40 mm × 40 mm × 2 mm displacement-amplifying compliant mechanism (DaCM) force sensor with optical calibration for the force feedback while grasping a cell with haptic interface. Lotti et al. [45] designed a compliant mechanism-based 2-DOF robot finger that is tendon operated. They showed the trajectory control results. Biagiotti et al. [46] described a preliminary stage of developed for an anthropomorphic robot hand with compliant mechanism where the module of the finger is designed with plastic and connected with elastic parts as hinge. Each pair of links forms an angle of 45°. It is guided by three tendons for each finger. The adduction–abduction is not present in this design. Linear motors are used for the actuation system. Strain gauges are used for joint positioning sensor, and tactile sensors are used for soft skin.

1.2.5 IPMC Applications in Space

Shahinpoor et al. [38] and Sahoo [40] reported a gripping device made using IPMCs, which can lift ten times of its weight with 5 V actuating voltage. The applications of IPMCs are dust wiper and wiper blade actuation, and they are tested for film moisture, electrolysis, vacuum and low temperature. (A sizeable displacement was still observed at −140°C.) It is also reported that at low frequencies, the IPMC offers a better displacement. (At 6 V, 0.1 Hz, it deflects 1.5 cm approx., IPMC dimensions—15 mm × 6 mm.) Bar-Cohen [4] and Steve Tung et al. [47] presented an overview on the applications of a MEMS-based flexible sensor and actuator system to monitor and control the health of an inflatable space structure. They developed two-dimensional pattern electrodes using IPMCs pasted in different parts of the sheet, and different actuation levels are applied in different parts. NASA [48] published a report on bone loss in space, and the IPMC can rectify this problem with high efficiency. Sahoo [40] reported a schematic representation of the robot lifting a glass bead and fibers with an array of actuators. Carpi et al. [49] reported the research areas using EAPs as follows – to improve/replace the damping actuator for boom or torns, second – mounting inside the boom EAP may useful and bending stiffness may be improved by this, shape control of double shell reflector might be performed by EAP actuator, finally shape control of large inflatable reflectors may be performed by interconnecting contracting helical EAP actuators. Kumar Krishen et al. [50] showed the concept of a tightly worn suit with IPMC devices in it. The research by NASA [48] suggested that the loss of bone mass can be minimized by vibrating body bones. Many sensor applications require vibration and precise pointing control mechanisms that can be enabled by the use of IPMCs. Incorporating IPMC-based artificial muscles in various parts of the spacesuit can provide countermeasure devices for reducing muscle atrophy, alleviating the bone loss of the astronauts and monitoring astronaut activities. Several electronic applications for intelligent information manipulation and data storing are also possible. Bhandari et al. [1] highlighted the fabrication process of IPMCs and reported the tip force after applying the voltage (3.5 V). This process is done after soaking the IPMC in different salt solutions. H^+, Li^+ and Cu^{++}, Li^+ ions show the best results regarding tip force. Adam, a wired reporter and freelance journalist from Oakland, reported few technical problems that astronauts come across in spacecraft, such as feet molting, bugs in space, bone and muscle loss, space blindness, solar superstorm and radiation, toxic dust and psychology in space. Kumar Krishen [51] reported on the space applications and suggested that IPMC technology can be envisioned for use in space robotics, human support, and vibration monitoring and control. Muhammad Farid et al. [52] in their brief review concluded that IPMCs are advantageous in biomimetic

Introduction to IPMCs and Applications

applications on land and underwater, and they are demonstrated for applications as a catheter system for human body and heart holding devices. They have been applied to actuate the fins in underwater robots and are used as a sonar system in underwater artificial robots. Andres Punning et al. [53] tested ionic polymers under X-ray radiation, gamma radiation, UV radiation, vacuum and low temperature (freezing) keeping reference samples in ambient conditions, and they concluded that the cosmic radiation does not impinge on the prolonged exploitation of ionic electroactive polymer (IEAP) actuators in space applications.

Bhattacharya et al. [10,54] reported on the fabrication of an IPMC actuator-based soft gripper and the characterization of IPMCs with different shapes and sizes. Chattaraj et al. [55] showed tip and shape estimation of IPMCs and developed a two-jaw IPMC-based gripper using hyper-redundant approximation. Bhattacharya et al. [56] developed an IPMC-based data glove capable of object identification.

1.3 SUMMARY

This chapter mainly focused on the understanding of IPMCs, their basic manufacturing processes and application potential in various fields of science and engineering. The electrical characterization of the IPMC is discussed with its electrical model and the control issues. IPMCs mostly show nonlinear behavior during their operation, so the control of IPMC with various techniques is one of the research areas that can be followed. This chapter also focused on various types of IPMC grippers for micro- to macro-manipulation, and compliant mechanism gives an added advantage to design such grippers. IPMCs represent a potential actuator under moisture with low actuation power. This phenomenon encourages the space researchers in future technology that can be the next research topic.

The next chapter will highlight the manufacturing processes of IPMCs and also the fabrication of IPMCs with different metal electrodes.

For new updated research opening and the methodology of performing those research works, follow Chapter 8.

In the next chapter, the reader will find how IPMCs are manufactured using different processes. For example, in Figure 2.4, a single-layered Ag-IPMC manufacturing process is described with block diagram.

REFERENCES

[1] Bhandari, B., Lee, G.-Y., Ahn, S.-H., 2012. A review on IPMC material as actuators and sensors: Fabrications, characteristics and applications, International Journal of Precision Engineering and Manufacturing Vol. 13, No. 1, pp: 141–163.

[2] Alici, G., Higgins, M. J., 2009. Normal stiffness calibration of microfabricated trilayer conducting polymer actuators, Smart Materials and Structures, Vol. 18, No. 6, pp: 1–10.

[3] Bar-Cohen, Y., Zhang, Q., 2008. Electroactive polymer actuators and sensors, MRS Bulletin, Vol. 33, No. 3, pp: 173–181.

[4] Bar-Cohen, Y., Bao, X., Stewart Sherrit, Lih, S., 2005. Characterization of the electromechanical properties of ionomeric polymer-metal composite (IPMC), Proceedings of the SPIE Smart Structures and Materials Symposium, EAPAD Conference, San Diego, CA, March, pp: 286–293.

[5] Bar-Cohen, Y., 2000. Electroactive polymers as artificial muscles-capabilities, potentials and challenges, In Handbook on Biomimetics, Y. Osada (Chief Ed.), Section 11, Chapter 8, "Motion" Paper-134, NTS Inc., pp: 1044–1052.

[6] Lughmani, W. A., Jho, J. Y., Lee, J. Y., Rhee, K., 2009. Modeling of bending behavior of IPMC beams using concentrated ion boundary layer, International Journal of Precision Engineering and Manufacturing, Vol. 10, No. 5, pp: 131–139.

[7] Bhattacharya, S., Chattaraj, R., Das, M., Patra, A., Bepari, B., Bhaumik, S., 2015. Simultaneous parametric optimization of IPMC actuator for compliant gripper, International Journal of Precision Engineering and Manufacturing, Springer, Vol. 16, No. 11, pp: 2289–2297.

[8] Kim, S. J., Lee, I. T., Kim, Y. H., 2007. Performance enhancement of IPMC actuator by plasma surface treatment, Smart Materials and Structures, Vol. 16, No. 1, pp: N6–N11.

[9] He, Q., Yu, M., Song, L., Ding, H., Zhang, X., Dai, D., 2011. Experimental study and model analysis of the performance of IPMC membranes with various thickness, Journal of Bionic Engineering, Vol.8, No.1, pp: 77–85.

[10] Bhattacharya, S., Bepari, B., Bhaumik, S., 2014. IPMC-actuated compliant mechanism-based multifunctional multifinger microgripper, Mechanics Based Design of Structures and Machines, Taylor & Francis, Vol. 42, No. 3, pp: 312–325.

[11] Maheswari, N., Reddy, A. N., Sahu, D. K., Ananthasuresh, G., 2009. Miniature compliant grippers with force-sensing, Proc. of 14th National Conference on Machines and Mechanisms, pp: 431–439.

[12] Paul, A., Dey, N., Bhattacharya, S., 2020. Similarity analysis of IPMC and EMG signal and comparative study of statistical features. Advancements in instrumentation and control in applied system applications, IGI Global, DOI: 10.4018/978-1-7998-2584-5.ch001, pp: 1–16.

[13] Bhattacharya, S., Halder, S., Sadhu, A., Banerjee, S., Sinha, S., Banerjee, S., Kundu, S., Bepari, B., Bhaumik, S., 2020. Characteristics of ionic polymer metal composite (IPMC) as EMG sensor, Advancements in Instrumentation and Control in Applied System Applications, IGI Global, DOI: 10.4018/978-1-7998-2584-5.ch006, pp: 98–107

[14] Brunetto, P., Caponetto, R., Dongola, G., Fortuna, L., Graziani, S., 2009. A Scalable Fractional Model for IPMC Actuator, PHYSCON, Catania, Italy.

[15] Leary, S., Bar-Cohen, Y., 1999. Electrical impedance of ionic polymeric metal composites, Proc. of SPIE's 6th Annual Int. Symposium on Smart Structures and Materials, 1–5 March, 1999, Newport Beach, CA, pp: 3669.

[16] Brunetto, P., Fortuna, L., Giannone, P., Graziani S., Strazzeri, S., 2009. Static and dynamic characterization of the temperature and humidity influence on IPMC actuators, IEEE Transactions on Instrumentation and Measurement, pp: 893–908.

[17] Kim, B., Kim, B. M., Ryu, J., Oh, I. H., Lee S. K., Cha, S. E., 2003. Analysis of mechanical characteristics of the ionic polymer metal composites (IPMC) actuator using cast ion-exchange film, Smart Structures and Materials Electroactive Polymer Actuators and Devices, Proceedings Vol. 5051, pp: 486–495.

[18] Wu, Y., Nasser S. N., 2004. Verification of Micromechanical Models of Actuation of Ionic Polymer-metal Composites (IPMCs), Smart Structures and Materials 2004: Electroactive Polymer Actuators and Devices (EAPAD), San Diego, CA, USA, Proc. SPIE 5385, p. 155.

Introduction to IPMCs and Applications

[19] Lee, M. J., Jung, S. H., Lee, S., Mun, M. S., Moon, I., 2006. Control of IPMC-based artificial muscle for myoelectric hand prosthesis, The First IEEE/RAS-EMBS Int. Conf. on Biomedical Robotics and Biomechatronics, pp: 1172–1177.

[20] Yu, M., Shen H., Dai, Z., 2006. Manufacture and Performance Investigation of Ionic Polymer Metal Composites (IPMC), Int. Joint Conf. of NABIO/SMEBA, pp: 376–381.

[21] Stoimenov, B. L., Rossitera J. M., Mukaia, T., 2007. Manufacturing of Ionic Polymer-metal Composites (IPMCs) that can Actuate into Complex Curves, Proceedings of SPIE - Vol. 6524, Electroactive Polymer Actuators and Devices (EAPAD), pp: 65240T.

[22] Porfiri, M., 2008. Charge dynamics in ionic polymer metal composites, Journal of Applied Physics, Vol. 104, No. 10, November, pp: 104915 (1–10).

[23] Chen, Z., Hedgepeth D. R., Tan, X., 2009. Nonlinear capacitance of ionic polymer – metal composites, Electroactive Polymer Actuator and Devices (EAPAD), pp: 728715-1–728715-12.

[24] Paquette, J. W., Kim, K. J., Kim D., Yim, W., 2005. The Behavior of Ionic Polymer–Metal Composites in a Multi-Layer Configuration, Institute of Physics Publishing Smart Materials and Structures Smart Material, Structure Vol. 14, pp: 881–888.

[25] Lopes, B., Branco, P. J. C., 2008. Electromechanical characterization of non-uniform charged ionic polymer-metal composites (IPMC) devices, 4th World Congress on Biomimetics, Artificial Muscles and Nano-Bio IOP Publishing Journal of Physics: Conf. Series 127–012004.

[26] Junga, K., Namb, J., Choia, H., 2003. Investigations on Actuation Characteristics of IPMC Artificial Muscle Actuator, Science Direct, Sensors and Actuators, pp: 183–192.

[27] Jain, R.K., Datta S., Majumder, S., 2012. Design and control of an EMG driven IPMC based artificial muscle finger, Chapter 15, Computational Intelligence in Electromyography Analysis: A Perspective on Current Applications and Future Challenges, pp: 362–390.

[28] Shan, Y., Leang, K. K., 2008. Application of feed-forward dynamics compensation in ionic-polymer metal composite actuators, Electroactive Polymer Actuators and Devices, Bar-Cohen, Y. (Ed.), Proc. of SPIE, Vol. 6927.

[29] Xue, D., Chen, Z., Hao, L., Xu, X., Liu, Y., 2010. Modeling and control of IPMC for micro-manipulation, Proc. of the 8th World Congress on Intelligent Control and Automation July 6–9 2010, Jinan, China, pp: 2401–2405.

[30] Richardson, R. C., Levesley, M. C., Brown, M. D., Hawkes, J. A., Watterson, K., Walker, P. G., 2003. Control of ionic polymer metal composites, IEEE/ASME Transactions on Mechatronics, Vol. 8, No. 2, June, pp: 245–253.

[31] Tsiakmakis, K., Brufau-Penella, J., Puig-Vidal, M., Laopoulos, T., 2009. A camera based method for the measurement of motion parameters of IPMC actuators, IEEE Transactions on Instrumentation and Measurement, Vol. 58, No. 8, August, pp: 2626–2633.

[32] Bonomo, C., Fortuna, L., Giannone P., Graziani, S., 2004. A sensor-actuator integrated system based on IPMCs, Proc. of IEEE, Vol. 1, pp: 489–492.

[33] Shahinpoor, M., Kim, K. J., 2005. Ionic polymer–metal composites: IV. Industrial and medical applications, Institute of Physics Publishing Smart Materials and Structures, Smart Materials and Structures, pp: 197–214.

[34] Feng, G. H., Chen, R. H., 2008. Fabrication and characterization of arbitrary shaped μIPMC transducers for accurately controlled biomedical applications, Science Direct, Sensors and Actuators, pp: 34–40.

[35] Jain, R. K., Datta, S., Majumder, S., Dutta, A., 2011. Two IPMC fingers based micro gripper for handling, International Journal of Advanced Robotic Systems, Vol. 8, No. 1, ISSN 1729-8806, pp: 1-9.
[36] Jain, R. K., Patkar, U. S., Majumdar, S., 2009. Micro gripper for micromanipulation using IPMCs (ionic polymer metal composites), Journal of Scientific & Industrial Research, Vol. 68, Jan., pp: 23-28.
[37] Lumia, R., Shahinpoor, M., 2008. IPMC Microgripper Research and Development, 4th World Congress on Biomimetics, Artificial Muscles and Nano-Bio IOP Publishing Journal of Physics, Conf. Series 127, 012002.
[38] Shahinpoor, M., Bar-Cohen, Y., Xue, T., Simpson, J.O., Smith, J., 1998. Ionic polymer-metal composites (IPMC) as biomimetic sensors and actuators, Proceedings of SPIE's 5th Annual Int. Symposium on Smart Structures and Materials, San Diego, CA, Proc. SPIE 3324, 251.
[39] Chen, Z., Shen, Y., Xi, N., Tan, X., 2007. Integrated sensing for ionic polymer-metal composite actuators using PVDF thin films, IOP Publishing, Smart Material and Structures, pp: S262-S271.
[40] Sahoo, H. K., Pavoor, T., Vancheeswaran, S., 2001. Actuators based on electroactive polymers, Current Science, Vol. 81, No. 7, 10, pp: 743-746.
[41] Deole, U., Lumia, R., 2006. Measuring the load-carrying capability of IPMC microgripper fingers, IEEE Industrial Electronics, IECON 2006 – 32nd Annual Conf., pp: 2933-2938.
[42] Ananthasuresh, G. K., Maheshwari, N., Reddy A. N., Sahu, D. K., 2010. Fabrication of spring steel and PDMS grippers for the micromanipulation of biological cells, Microfluidics and Microfabrication, Springer, pp: 333-354.
[43] Hegde, H., Ananthasuresh, G. K., 2009. Design of compliant mechanisms for practical applications using selection maps, 14th National Conf. on Machines and Mechanisms NIT, Durgapur, India, Dec., pp: 27-36.
[44] Rao, V. M., Ananthasuresh, G. K., 2009. Haptic feedback for injecting biological cells using miniature compliant mechanisms, 14th National Conf. on Machines and Mechanisms (NaCoMM09), NIT, Durgapur, India, December 17-18, pp: 181-186.
[45] Lotti, F., Tiezzi, P., Vassura G., Zucchelli, A., 2002. Mechanical structures for robotic hands based on the "compliant mechanism" concept, 7th ESA Workshop on Advanced Space Technologies for Robotics and Automation, Noordwijk, The Netherlands, Nov.
[46] Biagiotti, L., Lotti, F., Melchiorri C., Vassura, G., 2003. Mechatronic design of innovative fingers for anthropomorphic robot hand, Proceedings of IEEE Int. Conference on Robotics & Automation Taipei, Taiwan, Sept., 14-19, pp: 3187-3192.
[47] Tung, S., Witherspoon, S. R., Roe, L. A., Al Silano, Maynard, D. P., Ferraro, N., 2001. A MEMS-based flexible sensor and actuator system for space inflatable structures, Smart Materials and Structures 10, pp: 1230-1239.
[48] Good Vibrations, 2001, Science Mission Directorate, NASA.
[49] Carpi, F., Sommer-Larsen, P., de Rossi, D., Gaudenzi, P., Lampani, L., Campanile, F., 2005. Electroactive polymers: new materials for spacecraft structures, Proceedings of the European Conference on Spacecraft Structures, Materials and Mechanical Testing (ESA SP-581), Noordwijk, The Netherlands, 10-12 May 2005.
[50] Krishen, K., 2009. Space applications for ionic polymer-metal composite sensors, actuators, and artificial muscles, Acta Astronautica Vol. 64, Elsevier, pp: 1160-1166.
[51] Krishen, K., 2012. New technology innovations with potential for space applications, Acta Astronautica Vol. 64, Elsevier, pp: 324-333.

[52] Farid, M., Gang, Z., Khuong, T. L., Sun, Z. Z., Rehman, N. U., Rizwan, M., 2014. Biomimetic applications of ionic polymer metal composites (IPMC) actuators-a critical review, Journal of Biomimetics, Biomaterials and Biomedical Engineering, Vol. 20, pp: 1–10.
[53] Punning, A., Kim, K. J., Palmre, V., Vidal, F., Plesse, C., Festin, N., Maziz, A., Asaka, K., Sugino, T., Alici, G., Spinks, G., Wallace, G., Must, I., Poldsalu, I., Vunder, V., Temmer, R., Kruusamae, K., Torop, J., Kaasik, F., Rinne, P., Johanson, U., Peikolainen, A.-L., Tamm, T., Aabloo, A., 2014. Ionic electroactive polymer artificial muscles in space applications. Scientific Reports, Vol. 4, No. 6913, pp: 1–6. DOI: 10.1038/srep06913.
[54] Bhattacharya, S., Bepari, B., Bhaumik, S., 2016. Soft robotic finger fabrication with PDMS and IPMC actuator for gripping, IEEE Technically Sponsored SAI Computing Conference, 13–15 July in London, UK. pp: 403–408.
[55] Chattaraj, R., Khan, S., Bhattacharya, S., Bepari, B., Chatterjee, D., Bhaumik, S., 2016. Development of two jaw compliant gripper based on hyper - redundant approximation of IPMC actuators, Sensors & Actuators: A. Physical, Elsevier, Vol. 251, Nov., pp: 207–218.
[56] Bhattacharya, S., Das, R., Chakraborty, R., Dutta, T., Mondal, A., Sarkar, S., Bepari, B., Bhaumik, S., 2017. IPMC based data glove for object identification, 6th International Conference on Informatics, Electronics & Vision (ICIEV) & 7th International Symposium in Computational Medical and Health Technology (ISCMHT), University of Hyogo, Himeji, Hyogo, Japan, Sept. 1–3.

2 Ionic Polymer–Metal Composite Actuators
Methods of Preparation

Dillip Kumar Biswal
C.V. Raman Global University
Orissa, India

CONTENTS

2.1 Introduction 17
2.2 Actuation Mechanism 18
2.3 Literature Review 20
    2.3.1 Fabrication Techniques 20
    2.3.2 Chemical Decomposition Method 20
    2.3.3 Mechanical Plating Method 21
2.4 Methodology 22
    2.4.1 Fabrication of Single-Layered Ag-IPMC 23
    2.4.2 Fabrication of Multilayered Ag-IPMC 24
2.5 Result Discussion 25
    2.5.1 Characterization of Ag-IPMC 25
        2.5.1.1 Morphological and Microstructure Analysis 25
        2.5.1.2 Bending Characteristics of the Fabricated Ag-IPMC 25
2.6 Conclusions 27
References 28

2.1 INTRODUCTION

Electroactive polymers (EAPs), which have been discovered in the past few decades, are associated with energy exchanging smart materials and are greatly useful for various applications. Due to their low mass, softness, noiselessness and large deformation under a relatively low driving voltage (1 –3 V), during the recent period, ionic electroactive polymer actuators have extensively been studied by researchers as promising smart materials for their wide range of applications in various industries and academics, such as robotics, biomedical engineering and artificial muscles, and 3D printing [1–5].

Ionic polymer–metal composites (IPMCs) have several configurations, but they are commonly categorized with respect to the driving mechanism. Electronic EAPs are driven by an electric field governed by coulomb forces, while ionic EAPs

DOI: 10.1201/9781003204664-2

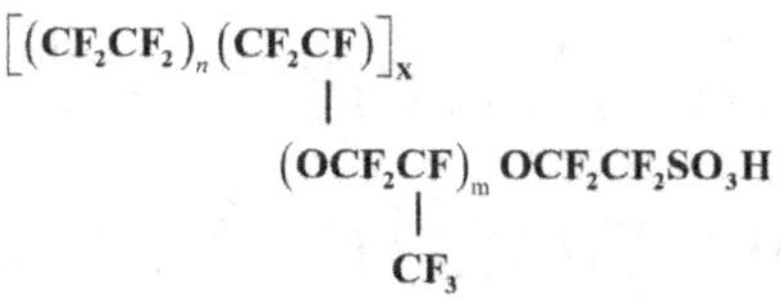

$$[(CF_2CF_2)_n(CF_2CF)]_x$$
$$|$$
$$(OCF_2CF)_m\,OCF_2CF_2SO_3H$$
$$|$$
$$CF_3$$

FIGURE 2.1 Structural formula of the Nafion polymer in acid form [9].

are driven by the mobility/diffusion of ions [7]. When a small electric potential of 1 V–3 V is applied across the thickness of an EAP material, it shows a considerable bending deflection, and it is also capable of producing a measurable electric potential when subjected to a mechanical deformation. These unique properties can be very useful in a variety of applications that require actuation or sensing [6–8].

One of the most widely used ionic polymers available today for the fabrication of IPMCs is DuPont™ Nafion®, which was developed in the early 1960s in collaboration with General Electric to be used as a fuel cell membrane [Shahinpoor, 2005]. The material is a Teflon-based polymer with sulphonic acid side groups. When the material is soaked in water at 100°C for 1 hour, it absorbs water as much as 38% of its dry weight, linearly expands as much as 15% of its original length and increases in thickness up to 14% [DuPont Product Information Sheet]. The chemical structure of the Nafion is shown in Figure 2.1 [9]. In the figure, m, n and X depend on the type of the polymer. For example, in Nafion-117 membrane, $m=1$, $n=6-7$, and $100 < X < 1000$. In fully hydrated condition, the ability to conduct cations across the thickness of the polymer membrane gives it the electromechanical transduction capability [10].

Although the ionic polymers have been commercialized in the early 1990s, their actuation and sensing capabilities have extensively been studied in the recent years. The use of Nafion-based composites as an electromechanical actuator was demonstrated by [11,12]. Both of them showed that applying an electric potential to the material results in a mechanical deformation. The sensing capability of IPMC under mechanical deformation by developing an accelerometer was demonstrated by Sadeghipour et al. [13]. These unique properties make ionic polymers an option in the field of active materials and make them suitable for being used as both actuators and sensors.

2.2 ACTUATION MECHANISM

As no established actuation mechanism for IPMC actuator is available across the world till date, the basic theory that manipulates the actuation mechanism of an ionic polymer is still under research. The electromechanical characteristics of ionic polymers were discovered over a decade ago [14–16]. The electromechanical coupling effect and changes in the physical phenomena complicate the actuation mechanism.

An IPMC consists of two basic components: a base polymer membrane (ionomer) and conducting metal electrodes, as shown in Figure 2.2. The main distinction between an IPMC actuator and a fuel cell is that a conductive electrode coating on

Methods of Preparation of IPMC Actuators

both sides of the ionomer surface is developed to facilitate the application of an electric potential that induces the bending deflection of the IPMC. Mostly, chemical or mechanical methods are employed to develop the metal electrode over the ionomer surfaces. Figure 2.2 shows the top and bottom surfaces representing the conducting electrode layers, where the outer surfaces are observed to be relatively flat and smooth. The interface between the electrode layer and the polymer is non-uniform due to the surface roughening and penetration of the electrode material into the surface of the membrane. The base material of the membrane is a perfluorinated ionomer that consists of a chemically and thermally stable fluorocarbon backbone, hydrophilic anion domains (fixed anion), hydrophobic side chains (mobile cations) and a solvent (water), as shown in Figure 2.3a. The most common solvent used is deionized water, which serves as a medium for ion migration within the base polymer. In neutral state, the mobile cations are bonded with the fixed anionic groups, forming a cluster with the solvent (water) molecules, and are also easily exchangeable with the other cations [17].

Some of the well-established, basic theories about the actuation mechanism are reviewed in this section. When an electric potential is applied, the weak ionic bond breaks apart as the positively charged cations are migrated towards the negatively charged cathode [18]. De-Gennes et al. [14] proposed a model that

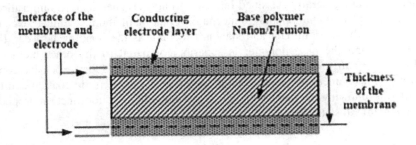

FIGURE 2.2 Representation of IPMC showing base polymer sandwiched between the metal electrodes on both sides of the surface.

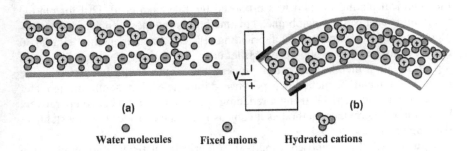

FIGURE 2.3 (a) Graphical representation of the IPMC actuator; (b) actuation mechanism under applied electric potential.

postulates water pressure gradients (mobile cations drag water molecules with them as they move towards the cathode, thus creating a pressure gradient) as the primary mechanism for actuation. Nemat-Nasser and Li [15] proposed a model that suggests that the ion transportation (redistribution of cations) causes the bending of IPMC.

There are plentiful mobile cations and solvent (water) molecules present over the ionic polymer membrane. When a small electric potential is applied across the thickness of the conducting electrode metal layer, mobile cations pooled together with the surrounding solvent molecules will move towards the oppositely charged electrode, i.e. the cathode side of the ionic polymer membrane. The migration of mobile cations initiate inequitable swelling near the cathode electrode and shrinkage near the anode electrode, which leads to bending towards the anode electrode, as shown in Figure 2.3b. However, it could be possible that both the forces actually play a role in the actuation mechanism of IPMC actuators.

2.3 LITERATURE REVIEW

2.3.1 Fabrication Techniques

The processes of developing a conductive electrode layer over the ionomeric membrane are generally classified into two groups: (1) chemical decomposition/electroless plating method and (2) mechanical plating method. Chemical decomposition is the predominant and standard method as the bonding between the electrode and the base polymer is superior compared to the other methods. However, the method is proved to be time-consuming and more expensive compared to mechanical methods. Even though a uniform electrode coating can be achieved following the mechanical plating method, it does not function properly in the hydrated state [19].

2.3.2 Chemical Decomposition Method

In the chemical decomposition/electroless plating method, a reducing agent is used for the reduction of metal particles inside the polymer membrane. Earlier, the chemical plating method was proposed by Takenaka et al. [20] for plating platinum over a Nafion membrane. This method is the most efficient electroless plating technique employed for fabricating IPMCs. Another method similar to the previous one was proposed by Millet et al. [21, 22] for plating platinum on top of a Nafion membrane. In this process, the membrane is immersed in the metal solution and the metal ion is penetrated into the Nafion membrane and later reduced to metal particles using a reducing agent. This method is comparatively better than the previous method as it can easily form electrodes that are stable in the long term.

Oguro et al. [23] proposed a method to fabricate an IPMC with gold as the surface electrode, where an ion exchange counterion such as Na^+ with a cationic (gold complex) solution is treated by a reduction process using reducing agents such as sodium sulphate. This ion exchange and reduction process can be repeated several times until a suitable thickness of the surface electrode is plated on the surface of the membrane.

Almost all the previous efforts to fabricate IPMCs used platinum or gold as the surface electrode, which contributes to the high cost of IPMCs restraining their probable applications in diverse fields. To trim down the cost of fabrication of IPMCs, Bennett and Leo [24] developed a co-reduction process with the combination of both precious and non-precious metal electrodes. The IPMC actuator fabricated by co-reducing copper and platinum (Cu–Pt) inside a Nafion membrane achieved a life cycle of 250,000 cycles (1.25 V, 1.0 Hz sine wave) in deionized water. To decrease the production cost, non-precious conducting metals such as silver were used [25].

2.3.3 Mechanical Plating Method

Although the fabrication of IPMCs using the chemical decomposition method produces mechanically stable electrodes, they are complex, expensive and time-consuming. Further, an effective fabrication requires a brief knowledge about the right/exact amount of chemical composition, and the reaction between the chemicals and the membrane. Hence, sometimes mechanical plating techniques are also preferred over chemical plating methods to fabricate IPMCs. Mainly three techniques are widely accepted for plating surface electrode over the ion exchange polymer membrane, i.e. physical vapour deposition, solution casting and direct assembly process.

E-beam evaporation, followed by electroplating two thick gold layers, was used to fabricate an IPMC [26]. The IPMC fabricated by this method can withstand a high voltage of the order of 20 V without peeling off of the electrodes. Sputter coating technique was used for depositing gold as a surface electrode on the Nafion surface [27]. The process proved to be fast and cost-effective for the fabrication of IPMCs compared to the chemical decomposition method. Both precious metals (gold and platinum) and non-precious metals (nickel, silver and copper) can be deposited as surface electrode without much change in the deposition technique. Lee et al. observed that an improvement in performance in terms of tip displacement and force output can be achieved for IPMCs fabricated by the chemical decomposition method, followed by sputter coating of gold over them [28]. However, cracks are formed on the surface of the IPMC as the sputter coating is performed in dry state.

The actuation force in an IPMC actuator is directly proportional to its dimensions and the applied electric potential. Even though a thick IPMC can develop a large tip force and can operate for a longer time, the bending tip deflection reduces considerably. For fabricating a thicker IPMC, solution casting and hot pressing techniques are generally followed. However, different shapes and sizes of IPMC transducers can be fabricated using solution casting methods.

Kim and Shahinpoor successfully developed IPMCs having a thickness greater than 2 mm from liquid Nafion with platinum as the surface electrode following the solution casting method [2]. Further, Nafion solution can be mixed with ion conducting powders such as platinum, gold, palladium, copper, silver and carbon for various electroding purposes. The ion conducting powder-coated electrode is cured under an elevated temperature for achieving stability.

The solution casting method was used by Chung et al. [29] to fabricate IPMCs with silver nano-powders. This IPMC showed a deformation of large bending curvature angle (more than 90°) with an input voltage of 3 V.

Similar to the physical vapour deposition method, surface electroding using solution casting suffers poor bonding between the polymer membrane and the electrode. Also, this method requires accurate adjustment of the process variables such as temperature and concentration of the solvents. Hence, the reproducibility may not be achieved exactly. To overcome this drawback, the hot pressing technique may be employed. In the hot pressing technique, a hot pressing system is used to integrate several thin ion exchange membranes of Nafion together. This method is expected to enhance the blocking force as the bending stiffness of the base polymer increases. The method is also proved to be efficient as far as repeatability is concerned.

A thick-film IPMC actuator was successfully fabricated by hot pressing several thin films of Nafion membrane, achieving a thickness between 0.54 and 0.9 mm [30]. Several cycles of Pt electrode plating were applied following the chemical decomposition method to improve the actuation performance of the IPMC actuator. Tiwari and Kim fabricated a disc-shaped IPMC using Nafion granules coated with conductive silver paint following the hot pressing technique [31].

Each fabrication process, i.e. chemical and mechanical plating methods, has its own advantages and disadvantages. This fact attributes that the choice of a particular method or a combination of different methods to fabricate IPMCs mainly depends on the stability and quality of the electrode, cost and the time consumed for fabrication.

However, there are very less commercial applications of IPMCs as actuators or sensors till date. One reason could be that the technology is still new and the working mechanism and further improvement on it is not well explored; another reason is that the manufacturing of IPMCs is very expensive. The reason behind the high manufacturing cost of these composites is due to the high cost of the precious metals such as platinum (Pt) or gold (Au) that are typically used as the surface electrode. The future potential applications of these polymers in various fields are remarkable, but are limited due to the high manufacturing cost. This motivates to use non-precious metals such as silver as the conducting surface electrode, which has been shown to be valuable for improving the performance of IPMC.

2.4 METHODOLOGY

The electroless plating method was followed to formulate the proposed Ag electrode IPMC actuator.

2.4.1 Fabrication of Single-Layered Ag-IPMC

The materials required for fabricating the Ag-IPMC are as follows:

- Nafion-117 as base polymer
- Silver nitrate (AgNO$_3$)
- Dilute ammonia solution (NH$_3$)
- Sodium hydroxide (NaOH)
- Dextrose anhydrous GR (C$_6$H$_{12}$O$_6$)
- Hydrochloric acid (HCl)
- Deionized/distilled water.

The step-by-step procedure to fabricate the Ag-IPMC is shown in the flow chart in Figure 2.4.

Pre-treatment: The Nafion-117 membrane surface is modified by roughening of both sides using a metallographic silicon carbide grinding paper. This process is performed to increase the surface area so that more silver ions can be deposited over the surface of the membrane and the depth of penetration will increase. After roughening, the Nafion-117 membrane goes through the chemical treatment process with HCl. In this process, impurities from the membrane are removed by boiling in hydrochloric acid (2N solution) for 30 minutes and by washing smoothly with distilled water. It is boiled in distilled water for over 30 minutes and kept in the distilled water till the commencement of the next step.

Adsorption: In this process, exchanging of Na$^+$ and Ag (NH$_3$)$_2$$^+$ ions into the Nafion-117 membrane is carried out. To carry out this process, a NaOH solution with the concentration of 0.50 mol/L is prepared. To ensure adequate diffusion of Na$^+$, the Nafion membrane is kept within the NaOH solution for 6–8 hours with continuous stirring for 1 hour. A diamminesilver(I) hydroxide solution, Ag(NH$_3$)$_2$OH, is prepared with a concentration of 0.15 mol/L. Then keeping

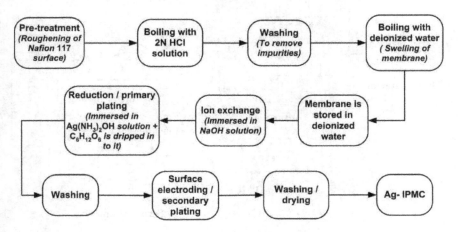

FIGURE 2.4 Flow chart to fabricate single-layered Ag-IPMC.

the membrane in Ag(NH₃)₂OH solution, it is stirred for 1 hour, and to diffuse Ag(NH₃)₂⁺ into the membrane, it is kept in the solution for 6 hours.

Reduction: A reducing agent solution ($C_6H_{12}O_6$) with the concentration of 0.0888 mol/L is prepared and mixed drop by drop into the solution of Ag(NH₃)₂OH, followed by stirring for few minutes. Equation (2.1) shows the reaction that takes place due to the addition of reducing agent to the diamminesilver(I) hydroxide solution [25]:

$$2Ag(NH_3)_2 OH + C_6H_{12}O_6 2Ag + C_6H_{11}O_2NH_4 + 3NH_3 + H_2O \quad (2.1)$$

The membrane is then placed in the NaOH solution, which is stirred in the ultrasonic cleaning bath for 30 minutes at 20°C–25°C to decontaminate any slackly attached silver particles from the surface of the membrane. The membrane is then stored in the distilled water till the next stage is performed.

Developing: In this step, the membrane is immersed deep into the $C_6H_{12}O_6$ solution at room temperature for approximately 2 hours. Then for approximately 3 hours, the membrane is further soaked in the diamminesilver(I) hydroxide solution at room temperature. With continuous stirring, the $C_6H_{12}O_6$ solution is mixed drop by drop into the Ag(NH₃)₂OH solution that contains the membrane. During this process, a silver layer is produced over the surface of the membrane. The membrane is then rinsed with distilled water and stirred in the ultrasonic cleaning bath to remove any loosely deposited silver particles on the membrane surface.

2.4.2 Fabrication of Multilayered Ag-IPMC

The multilayered coating of silver particles was carried out by repeating the reduction process over the conventional fabrication process of Ag-IPMC, which is illustrated in Figure 2.5.

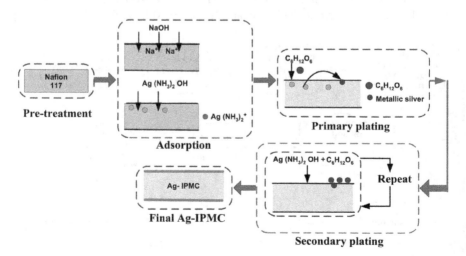

FIGURE 2.5 Fabrication steps for multilayered IPMC actuator.

Methods of Preparation of IPMC Actuators

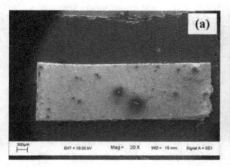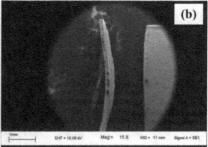

FIGURE 2.6 Images of the fabricated single-layered Ag-IPMC actuator: (a) top view; (b) side view.

Figure 2.6a and b shows the top and side surface images of the fabricated IPMC taken using a scanning electron microscope (SEM) with the amplification ratios of 20× and 15×, respectively.

2.5 RESULT DISCUSSION

2.5.1 Characterization of Ag-IPMC

2.5.1.1 Morphological and Microstructure Analysis

A SIGMA field emission SEM is used to study the microstructure of the fabricated Ag-IPMC. Figure 2.7a shows the SEM-based morphological analysis of the fabricated single-layered Ag-IPMC. The micrograph shows tightly filled and satisfactory spreading of silver particles over the surface of the IPMC. Figure 2.7b demonstrates the chemical composition of the sample in its surface. The result clearly shows the presence of Ag particles over the surface of the IPMC.

From Figure 2.8a, it is clearly observed that the silver particles go through the surface of the Nafion membrane up to 8–9 μm, and Figure 2.8b reveals the mixture of Ag particles and Nafion membrane compounds present over the cross section of the IPMC. The morphological study reveals that the average diameter of the Ag particle is around 0.5–0.6 μm.

2.5.1.2 Bending Characteristics of the Fabricated Ag-IPMC

Single-layered and multilayered silver electrode IPMC actuators were fabricated and cut into a size of 20×5×0.2 (mm^3) for the experimental study. A conventional Ag-IPMC, a three-layered Ag-IPMC and a four-layered Ag-IPMC were fabricated by taking the same chemical composition and following the chemical decomposition method for the entire specimen. The experiment was conducted in fully hydrated condition of the IPMC actuator in a fixed–free mode. The actuator is fixed at one end where copper strips are used to provide the controlled voltage (0.2–1.2 V) quasi-statically from a DC power source (0–32 V DC, 0–2 A, Testronix). The other end of the actuator is free. During this process, the bending deformation of the IPMC actuator is observed.

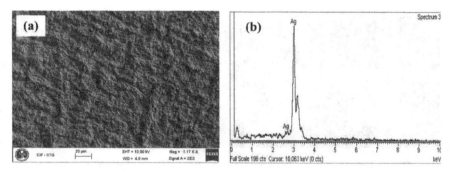

FIGURE 2.7 (a) SEM micrograph; (b) EDX of the surface of the fabricated IPMC.

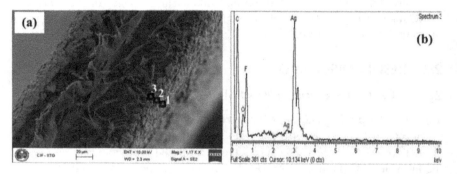

FIGURE 2.8 (a) SEM micrograph; (b) EDX of the cross section of the fabricated IPMC.

The graphical representation of bending configuration is shown in Figure 2.9a, while Figure 2.9b shows the experimental set-up and the bending actuation of single-layered Ag-IPMC actuator subjected to a DC input voltage of 1.2 V.

For each input voltage, after 30 seconds the tip position (p_x, p_y) of the IPMC actuator has been considered and the experimental data for the tip position (p_x, p_y) with an applied voltage (V) for the first, third and fourth coatings are plotted in Figure 2.10a and b, respectively. From the graph, it is clearly observed that as the thickness of coating over the base material (Nafion-117) increases, the value of X-coordinates decreases and that of Y-coordinates increases; i.e., the tip deflection increases with the increase in thickness.

A theoretical relationship has been established by considering the experimental X-coordinate and Y-coordinate tip deflection data. The experimental tip deflection data for various input voltages are plotted for the first, third and fourth coatings with a cubic-order polynomial curve fitting approximation as shown in Figure 2.11a–c. It is clearly observed from the graph that the pattern of bending is almost the same for all the Ag-IPMC actuators.

Methods of Preparation of IPMC Actuators

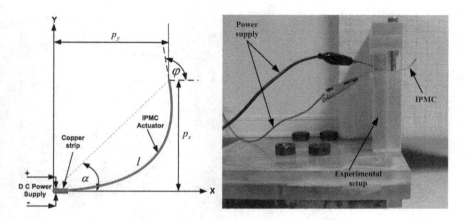

FIGURE 2.9 (a) Graphical representation of the deflected pattern of an IPMC actuator for an applied voltage; (b) effect of single-layered IPMC under applied electric potential of 1.2 V DC.

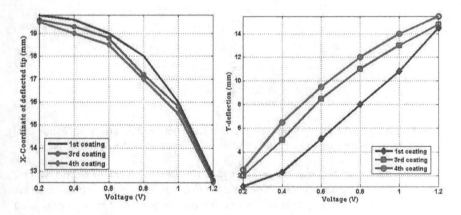

FIGURE 2.10 Tip position in terms of the (a) X-coordinate and (b) Y-coordinate of the Ag-IPMC actuator for different values of the input voltage.

2.6 CONCLUSIONS

In outline, the electroless plating method has been followed to fabricate single-layered and multilayered silver-coated ionic polymer–metal composites (IPMCs). Compared to the usual single-layered Ag-IPMC actuator, a notable improvement in bending deformation is observed in the multilayered Ag-IPMC actuator. This study would be helpful for the development of high-performance IPMC actuators and could be useful in applications where a higher actuation force at a higher bending displacement is required, such as small-scale robots and underwater vehicles.

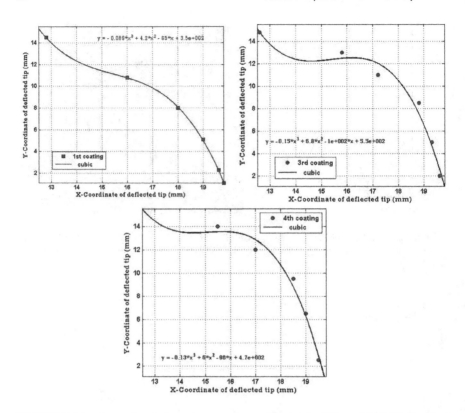

FIGURE 2.11 Curve fitted along the experimentally obtained X- and Y-deflection data for the (a) first coating, (b) third coating and (c) fourth coating of the Ag-IPMC actuator.

In the next chapter, the reader will find the new concept of ocean kinetic energy harvesting process. The process of activation for complex IPMC systems and wave amplitude are physically modelled and validated. This is a new area for the researchers.

REFERENCES

[1] Shahinpoor, M., and Kim, K.J. (2005). Ionic polymer-metal composites: IV. Industrial and medical applications. Smart Materials and Structures, **14**, 197–214.

[2] Nemat-Nasser, S., and Li, J.Y. (2000). Electromechanical response of ionic polymer-metal composites. Journal of Applied Physics, **87** (7), 3321–3331.

[3] Del Bakhshayesh, A.R., Asadi, N., Alihemmati, A., Nasrabadi, H.T., Montaseri, A., Davaran, S., and Abedelahi, A. (2019). An overview of advanced biocompatible and biomimetic materials for creation of replacement structures in the musculoskeletal systems: focusing on cartilage tissue engineering. Journal of Biological Engineering, **13** (1), 85.

[4] Hosseini, V., Maroufi, N.F., Saghati, S., Asadi, N., Darabi, M., Ahmad, S.N.S., and Rahbarghazi, R. (2019). Current progress in hepatic tissue regeneration by tissue engineering. Journal of Translational Medicine, **17** (1), 383.

Methods of Preparation of IPMC Actuators

[5] Carrico, J.D., Hermans, T., Kim, K.J., Leang, K.K. (2019). 3-D-Printing and machine learning control of soft ionic polymer-metal composite actuator. Scientific Reports, **9**, 17482. https://doi.org/10.1038/s41598-019-53570-y.

[6] Kim, K.J. and Shahinpoor. M. (2002). A novel method of manufacturing three-dimensional ionic polymer–metal composites (IPMCs) biomimetic sensors: actuators and artificial muscles. Polymer, **43** (3), 797–802.

[7] Kim, K.J. and Shahinpoor, M. (2003). Ionic polymer—metal composites: II. Manufacturing techniques. Smart Materials and Structures, **12** (1), 65.

[8] Punning, A., Kruusmaa, M. and Aabloo, A. (2007). A self-sensing ion conducting polymer metal composite (IPMC) actuator. Sensors and Actuators A: Physical, **136** (2), 656–664.

[9] Seen, A.J. (2001). Nafion: an excellent support for metal-complex catalysts. Journal of Molecular Catalysis A: Chemical, **177**, 105–112.

[10] Newbury, K., and Leo, D.J. (2002). Electromechanical modeling and characterization of ionic polymer benders. Journal of Intelligent Material System and Structures, **13**, 51–60.

[11] Oguro, K., Kawami, Y., and Takenaka, H. (1992). Bending of an ion-conducting polymer film electrode composite by an electric stimulus at low voltage. Journal of Micro-Machine Society, **5**, 27–30.

[12] Segalman, D., Witkowski, W., Adolf, D., and Shahonpoor, M. (1992). Theory of electrically controlled polymeric muscles as active materials in adaptive structures. Smart Materials and Structures, **1**, 44–54.

[13] Sadeghipour, K., Salomon, R., and Neogi, S. (1992). Development of a novel electrochemically active membrane and 'smart' material based vibration sensor/damper. Smart Materials and Structures, **1**, 172–179.

[14] De-Gennes, P.G., Okumura, K., Shahinpoor, M., and Kim, K.J. (2000). Mechanoelectric effects in ionic gels. Europhysics Letters, **50**, 513–518.

[15] Nemat-Nasser, S., and Li, J. (2000). Electromechanical response of ionic polymer-metal composites. Journal of Applied Physics, **87**, 3321–3331.

[16] Tadokoro, S., Yamagami, S., Takamori, T., and Oguro, K. (2000). Modeling of Nafion-Pt composite actuators (ICPF) by ionic motion. SPIE Smart Structures and Materials, San Diego, CA, **3987**, 92–102.

[17] Lopez, M., Kipling, B., and Yeager, H. (1977). Ionic diffusion and selectivity of a cation exchange membrane in nonaqueous solvents. Analytical Chemistry, **49**, 629–632.

[18] Lakshminarayanaiah, N. (1969). Transport Phenomena in Membranes, Academic Press, New York.

[19] Tiwari, R., and Garcia, E. (2011). The state of understanding of ionic polymer metal composite architecture: a review. Smart Materials and Structures, **20** (8), 083001 DOI: 10.1088/0964–1726/20/8/083001.

[20] Takenaka, H., Torikai, E., Kawami, Y., and Wakabayashi, N. (1982). Solid polymer electrolyte water electrolysis. International Journal of Hydrogen Energy, **7**, 397–403.

[21] Millet, P., Pineri, M., and Durand, R. (1989). New solid polymer electrolyte composites for water electrolysis. Journal of Applied Electrochemistry, **19**, 162–166.

[22] Millet, P., Pineri, M., and Durand, R. (1995). Preparation of solid polymer electrolyte composites: investigation of the precipitation process. Journal of Applied Electrochemistry, **25**, 233–239.

[23] Oguro. K., Fujiwara, N., Asaka, K., Onishi, K., and Sewa, S. (1999). Polymer electrolyte actuator with gold electrodes. SPIE Conference on Smart Structures and Materials, **3669**, 64–71.

[24] Bennett, M.D., and Leo, D.J. (2003). Manufacture and characterization of ionic polymer transducers employing non-precious metal electrodes. Smart Materials and Structures, **12**, 424–436.

[25] Chen, Q., Xiong, K., Bian, K., Jin, N., and Wang, B. (2009). Preparation and performance of soft actuator based on IPMC with silver electrodes. Frontiers of Mechanical Engineering in China, **4**, 436–440.

[26] Zhou, W., Li, W.J., Xi, N., and Ma, S. (2001). Development of force-feedback controlled Nafion micromanipulators. Proceeding of Smart Structures and Materials, **4329**, 401–410.

[27] Siripong, M., Fredholm, S., Nguyen, Q.A., Shih, B., Itescu, J., and Stolk, J. (2006). A cost-effective fabrication method for ionic polymer-metal composites. Proceeding Material Research Society Symposium, **889**, 139–144.

[28] Lee, S-G., Park, H-C., Pandita, S.D., and Yoo, Y. (2006a). Performance improvement of IPMC (ionic polymer metal composites) for a flapping actuator. International Journal of Control, Automation, and Systems, **4**, 748–755.

[29] Chung, C.K., Fung, P.K., Hong, Y.Z., Ju, M.S., Lin, C.C.K., and Wu, T.C. (2006). A novel fabrication of ionic polymer-metal composites (IPMC) actuator with silver nano-powders. Sensors and Actuators B, **117**, 367–375.

[30] Lee, S.J., Han, M.J., Kim, S.J., Jho, J.Y., Lee, H.Y., and Kim, Y.H., (2006b). A new fabrication method for IPMC actuators and application to artificial fingers. Smart Materials and Structures, **15**, 1217–1224.

[31] Tiwari, R., and Kim, K.J. (2010). Disc-shaped ionic polymer metal composites for use in mechano-electrical applications, Smart Materials and Structures, **19** (6), 065016 DOI: 10.1088/0964–1726/19/6/065016.

3 A Study of Movement, Structural Stability, and Electrical Performance of a Harvesting System Based on Ionic Polymer–Metal Composites

Nang Xuan Ho and
Vinh Nguyen Duy
Phenikaa University

Hyung-Man Kim
Inje University

CONTENTS

3.1 Introduction 32
3.2 Modeling of the IPMC Movement and Ocean Wave Kinetic Energy 35
    3.2.1 Modeling of the Relationship between the Input Bending Angle and the Output Voltage 35
    3.2.2 Fabrication of IPMC 38
    3.2.3 Ocean Environmental Conditions 39
    3.2.4 Simulation and Experimental Setup 40
3.3 Results and Discussion 43
    3.3.1 The Pressure and Motion Results 43
    3.3.2 Effect of Wave Direction and Frequency on the Movement of the Modules 44
    3.3.3 Effect of the Wetted Surface Area and Mass on the Movement of the Modules 48

DOI: 10.1201/9781003204664-3

3.3.4 Performance Test of the Ocean Kinetic Energy-Harvesting Modules 49
3.4 Conclusions 49
References 51

3.1 INTRODUCTION

In recent years, the world's growing demand for energy has been a great challenge, and sustainable development needs to be achieved; therefore, it is necessary to find alternative energy sources because fossil fuels such as coal and petroleum fail us in many ways. Among all renewable energy sources, ocean waves have the highest energy density and are considered by scientists worldwide as one of the potential renewable energy sources to support humankind [1–3]. Studies on the production of usable energy from the ocean's kinetic energy have motivated scientists worldwide since the late 20th century, such as the studies of wave energy converters (WECs) published worldwide and other different technologies for extracting energy from marine currents [4–11].

Ocean wave energy may be transformed into electricity by devices or power absorbers manufactured explicitly for this purpose. Piezoelectric materials have already been utilized to convert kinetic energy into electrical energy and have played an important role in electrical energy generation in several applications [12]. For example, the characteristics of the piezoelectric material, macro-fiber composites (MFCs), were investigated by Youngsu Cha [13]. This research analyzed energy harvesting from underwater vibrations of a piezoelectric structure composed of a thin aluminum beam sandwiched by two MFCs and focused on harmonic base excitation to explore the role of the water immersion depth and the amplitude and frequency of excitation in underwater vibrations. Experiments are also conducted to identify the piezoelectric properties and offer further insights into hydrodynamic forcing's role in structural vibrations. However, as mentioned in the literature, piezoelectric materials are brittle and do not respond actively at low frequencies, which has limited their applications [12]. This restriction can be minimized by using ionic polymer–metal composite (IPMC) materials due to their advantages of durability, frequency response, and directional response [12]. IPMC materials, a type of smart materials, have been initially used in various biomedical applications. IPMC materials are electroactive polymers made from synthetic composite materials. IPMCs have already been applied in actuators and sensors, and they behave similar to biological muscles when subjected to an electric field [14,15]. When the IPMC is bent due to kinetic energy, an output voltage proportional to the amplitude and direction of displacement is generated between the two electrodes across the membrane in both air and water.

The instantaneous power density of the state-of-the-art IPMC is approximately 20 W/cm^2, with an average efficiency of less than 2%. Inspired by the hydrophilic capability of the IPMC, some research topics challenged the development of a method where sensor-level powered IPMC materials can be scaled up to a portable power generator [15]. In another application, as presented by Janusz Kwaśniewsk [14], the results of preliminary studies concerning the creation of a precise mathematical description of the IPMC element were presented. Such a model could simplify the design and prototyping of a new generation of actuators, sensors, and energy harvesters. This research concerning the energy-harvesting possibilities was connected to the self-excited acoustical system.

In this research, two IPMC samples connected in series and parallel were investigated to determine their influence on each other. The results showed that these systems could restore stationary battery chargers and reduce the electrical power grid. Also, energy harvesting from underwater torsional vibrations of a patterned IPMC was investigated by Youngsu Cha [13]. This study focused on harmonic base excitation of a centimeter-sized IPMC, which was modeled as a slender beam with a thin cross section vibrating in a viscous fluid. Model parameters were identified from the in-air transient response, underwater steady-state vibrations, and electrical discharge experiments. The resulting electromechanical model allows predicting energy harvesting from the IPMC as a function of the shunting resistance and the base excitation frequency and amplitude.

In a previous study [16], we investigated an IPMC sensor model using an RC circuit in which the electrical components were related to the physical parameters of the IPMC. A charging model describing the linear kinematic relationship between the charge distributions and the applied bending angles was developed, and the time derivative of the charge model was implemented in the circuit model as a source of current. This research's main contribution was creating a practical IPMC sensor model that is readily applicable to the real world. Another contribution was that the IPMC sensor's performance had been demonstrated in an actual biomedical application, which indicated that further biomedical applications of the IPMC sensor were possible. Besides, to generate electricity from the ocean's kinetic energy, we developed the world's first movable ocean kinetic energy-harvesting module with an electrochemical material consisting of IPMC, which replaces fixed tidal power generators in driving turbines using the ocean flow [11]. Ultimately, we constructed a movable module for stand-alone ocean wave energy potential such that the power grid is out of reach to accommodate situations that require reduced costs and expanded coverage. During our experiments with IPMC, we encountered an unexpected growth of algae and barnacles that affected the electrochemical conversion.

IPMCs, consisting of two metal electrodes and an ion-conducting polymer, are promising classes of ionic electroactive polymers. They can be utilized as sensors, actuators, or energy harvesters [17]. Generally, perfluorinated polymers, such as sulfonated polymers (Nafion) or carboxylated polymers (Flemion), are employed for IPMCs [17]. Perfluorinated sulfonic acid ionomeric polymers are synthesized by copolymerization of sulfonyl fluoride vinyl ether and tetrafluoroethylene [18,19]. Nafion is a perfluorinated sulfonic acid ionomer membrane with a Teflon-like backbone and short side chains terminated by the sulfonic acid group, with counterions, such as H+, Li+, Na+, and K+, and hydrophobic fluorocarbon and hydrophilic ionic phases. When the IPMC is bent mechanically, the hydrated cations on the compressed side of the membrane move toward the membrane's stretched side, resulting in the imbalance in the number of cations contacting each electrode, and this produces an output voltage across the membrane.

The IPMC has been studied and developed as an actuator for application as a biomechanical device, a biomimetic device, or an artificial muscle actuator. It can be used as a catheter for a blood vessel surgery or a peristaltic micropump and robot fish. However, the IPMC actuator is required to produce more generative force to use it in various applications. When an electric field is applied to the metallic electrodes on an IPMC, the electrostatic force attracts the hydrated cations inside the membrane toward the boundary layer that is contacting the cathode, while anions are still fixed in the backbone of the polymer. The IPMC actuator's performance is dramatically affected by factors such as the type of inner solvent, cations, and electrode materials. The surface of the electrode plays an essential role in the actuation performance of IPMC. Typically, platinum is used for the electrode; however, the bending motion of IPMC can create cracks on the electrode's surface. Therefore, silver and gold particles can be added to the surface electrode to reduce stiffness and increase surface conductivity.

IPMC has also been used as a sensor since the IPMC films produce electromotive voltage when bending or being deformed. The IPMC sensor's output had a quasi-static relationship to displacement, and the IPMC could be used as a motion sensor or pressure sensor. When the IPMC is bent, the hydrated cations in the compressed side of the membrane will be forced to migrate toward the membrane's stretched side. Consequently, a relatively higher amount of hydrated cations will be accumulated on the stretched side of the electrode, while a deficit of cations occurs on the compressed side of the electrode. The imbalanced number of cations contacting each electrode causes the polarization of the IPMC across the membrane, generating a voltage that can be detected through surface electrodes. The essential factor in the performance of the IPMC is also the amount of inner solvent. Also, the surface electrode conductivity is crucial in the performance of the IPMC sensor. IPMC sensors with surface electrodes of high conductivity generate more noiseless output signals [26].

IPMCs are also promising in the field of energy harvesting. Energy harvesting utilizing smart materials can be a breakthrough in energy saving. Thus, it can provide some additional environmental protection. Such materials enable the creation of portable systems that can scavenge energy from the environment [21]. When an IPMC bends due to kinetic energy, a voltage is generated between the two electrodes across the membrane in both air and water. The instantaneous power density of the state-of-the-art IPMC is approximately 20 W/cm^3, with an average efficiency of about 2% or less. In some previous studies, IPMCs were applied to generate electricity from the kinetic energy in the ocean; these studies aimed to develop the world's first movable ocean kinetic energy-harvesting module with an electrochemical material consisting of IPMC that replaces fixed tidal power generation to drive turbines using ocean flow [11,21]. Consequently, they constructed a movable module for stand-alone ocean wave energy potential such that the power grid is out of reach to accommodate situations that require reduced costs and expanded coverage.

Based on the results of the above studies, this research designed a portable power system utilizing vertical waves and horizontal ocean currents using the electrochemical material consisting of IPMC to supply electricity to stand-alone offshore plants. Consequently, the ocean kinetic energy-harvesting system comprising a main buoy and 18 modules of attached IPMC materials was investigated to evaluate its energy-harvesting ability. Our goal is to develop a high-performance energy-harvesting system that coexists with the ocean, a complex three-dimensional world, and would advance marine resources conservation and development. For that reason, numerical studies of the energy-harvesting system using IPMC materials operating in the ocean environment were conducted with the 3D simulation software to evaluate its movement and structural stability and identify the optimal setup to enhance its energy-harvesting performance. Also, experimental work was undertaken to appraise the performance of the IPMC system. The results showed that the IPMC system's movement is generally stable and depends mainly on the incident wave. However, the amplitude of each module fluctuates dramatically concerning the resonant frequency. Also, the average power density at a given time of the day intensely oscillated above 180 $\mu W/m^2$ and reached a peak value of 280 $\mu W/m^2$. Each day's average power density was stable, approximately 245 $\mu W/m^2$ for more than two months. This project's results are the foundation to assess the ocean wave energy potential and the ability to use the developed setup for offshore energy harvesting to meet the global energy demand.

3.2 MODELING OF THE IPMC MOVEMENT AND OCEAN WAVE KINETIC ENERGY

3.2.1 Modeling of the Relationship between the Input Bending Angle and the Output Voltage

As mentioned in [20,21], the IPMC material is made from a polyelectrolyte membrane, including cations with a solvent and metal electrodes chemically retrofitted on both surfaces of the membrane, as shown in Figure 3.1. Meanwhile, SO_3^- ions are attached on the two electrode sides. As a result, it is fixed, but the cations and solvent can transfer inside the membrane. Under wet conditions where IPMCs are submerged in ocean water, the cations are covered by water molecules because of the electrostatic force. When the IPMCs are bent under the impact of ocean forcing, an imbalance in the electricity is created between the two electrodes, and this phenomenon contributes to generating the output voltage, as is clearly described in Figure 3.1.

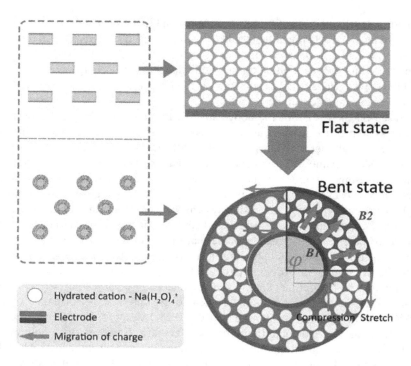

FIGURE 3.1 Electricity generation of ionic polymer–metal composites (IPMCs) under force impact.

To model the electricity generation process of the IPMC material, it is confirmed that the number of cations of the IPMC material is converted during the bending process. Figure 3.1 shows the connection between the net charges (q) and bending angles (φ_B). Accordingly, there is a difference in the hydrated cations between the two sides of the electrode when the IPMC is bent. It generates a difficult voltage between the two electrodes of the IPMC, and the net charge (q) is proportional to the angle (φ_B) or the area (ΔS) difference between the stretched and compressed sides. As shown in Figure 3.1, the expression for the difference of area (ΔS) can be calculated as follows:

$$\Delta S = b(B_1 - B_2) = b\left[(r+t)\frac{\pi\varphi_B}{180} - r\frac{\pi\varphi_B}{180}\right] = bt\frac{\pi\varphi_B}{180} \tag{3.1}$$

where r is the curvature radius; t is the membrane thickness; $\alpha\varphi_B$ is the arc angle (α is the coupling constant between the angles); B_2 and B_1 are the lengths of the inside and outside electrodes, respectively; and b is the IPMC's width. It is assumed that the net charge (q) generated by the bending process is proportional to the area (ΔS), and thus:

$$q \approx K\varphi_B \tag{3.2}$$

Energy Harvesting System Based on IPMCs

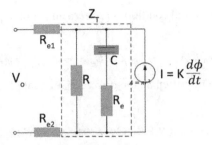

FIGURE 3.2 Circuit model of the IPMC material.

where K is an aggregated constant. Accordingly, we can calculate the current (I) generation as follows:

$$I = K\frac{d\varphi_B}{dt} \qquad (3.3)$$

We can model the output voltage of the IPMC material during the bending process, as shown in Figure 3.2. The IPMC material is connected to the measurement circuit via R_{e1} and R_{e2} resistances; meanwhile, R_c and R are the resistance of ion diffusion and the IPMC resistance, respectively. Because the resistances R_{e1} and R_{e2} are trivial compared to the polymer resistance, the output voltage (V_o) can thus be expressed by the following:

$$V_o = I Z_T \qquad (3.4)$$

where Z_T is the equivalent resistance of the circuit and can be expressed by Laplace domain expression as follows [21]:

$$Z_T = \frac{sRR_cC + R}{s(R+R_c)C+1} \qquad (3.5)$$

Then, the output voltage (V_o) is written by

$$V_{o(s)} = IZ_{T(s)} = K\frac{d\varphi_{B(t)}}{d_t}\frac{sRR_cC + R}{s(R+R_c)C+1} \qquad (3.6)$$

Applying the forward Laplace transform: $I_{(s)} \rightarrow K.s.\varphi_{B(s)}$ to Eq. (3.6):

$$V_{o(s)} = IZ_{T(s)} = Ks\varphi_{B(s)}\frac{sRR_cC + R}{s(R+R_c)C+1} \qquad (3.7)$$

$$\Leftrightarrow V_{o(s)}[s(R+R_c)C+1] = Ks\varphi_{B(s)}(sRR_cC + R) \qquad (3.8)$$

$$\Leftrightarrow V_{o(s)} + V_{o(s)}s(R+R_c)C = KRs\varphi_{B(s)} + KRR_cCs^2\varphi_{B(s)} \qquad (3.9)$$

After assuming all the initial conditions are zero and applying the inverse Laplace transform:

$$sV_{o(s)} \to V'_{o(t)};\ V_{o(s)} \to V_{o(t)};\ S^2\varphi_{B(s)} \to \varphi''_{B(t)};\ S\varphi_{B(s)} \to \varphi'_{B(t)}$$

Equation (3.9) can be rewritten as the differential equation of the output voltage as follows:

$$V_{o(t)} + V'_{o(t)}(R+R_c)C = KR\varphi'_{B(t)} + KRR_cC\varphi''_{B(t)} \qquad (3.10)$$

$$\Leftrightarrow V'_{o(t)} + \frac{1}{(R+R_c)C}V_{o(t)} = \frac{KRR_c}{R+R_c}\varphi''_{B(t)} + \frac{KT}{(R+R_c)C}\varphi'_{B(t)} \qquad (3.11)$$

3.2.2 Fabrication of IPMC

Because the current manufacturing processes of IPMCs are based on expensive noble metal platinum complex solutions, the application of IPMCs for the sensor, actuator, and energy harvesting has been limited to the research environment. Therefore, various studies have been investigated to find the best method for more efficient fabrication. These studies tried to replace expensive noble metal platinum and improve existing elaborate electroless plating. As described in [26], an IPMC was fabricated using a Nafion® 117 membrane with a typical thickness of 178 μm and a platinum ammine complex. The particles were plated on both sides of the membrane by the electroless plating method. The fabrication procedure has three main steps: pre-treatment of the polymer membrane, ion exchange process, and platinum particle plating. The pre-treatment of the polymer membrane makes Nafion® 117 polymer membrane suitable for electroless plating and maximizes the interfacial area between the membrane and the metal. In this process, the membrane is cleaned and fully hydrated where the anions at the backbone of the polymer combine with H+ ions. The ion exchange process is for adsorbing platinum complex on the surface of Nafion® membrane by dispersing platinum complex and by exchanging H+ ion with $[Pt(NH_3)_4]^{2+}$ ion. Meanwhile, plating of the platinum particle is for the plate on the membrane's surface using oxidation and reduction responses. After the plating process is finished, the fabricated IPMC was cleaned with deionized water for the experiment. The IPMC strips are kept in the chloride salt solution so that the cations in the membrane are exchanged with Na+.

The IPMC is one of the electroactive polymers and is known as a smart material; thus, it has been applied in various fields. However, as mentioned above, its application is limited since the production cost of IPMC is very high. Therefore, studies for establishing IPMC characteristics such as shape, size, material, and environment parameters to optimize the IPMC performance are necessary. IPMCs can generate a large bending motion under relatively low input voltage. However, conventional water-based IPMCs significantly lose their solvent content when operated by a voltage higher than 1.22 V in the air. This fact should be overcome for various potential applications such as artificial muscle actuators for insect-mimicking flapping devices. To prevent solvent loss, the replacement of the water solvent with ionic liquid has been studied [27]. Generally, the electric energy generation of IPMCs is due to the migration of hydrated cations, and they have a higher performance in water than in air. The effect of surrounding environments such as water, air, and the solution containing NaCl on the IPMC efficiency is described in [12]; it is shown that the most efficient performance of the IPMC was obtained when it was operated in a solution containing NaCl. Also, the IPMC performance directly depends on the IPMC's thickness; an increase in the thickness of the IPMC decreases the performance of the IPMC, as described in [20].

3.2.3 Ocean Environmental Conditions

Ocean waves are composed of waves of different frequencies and directions. The ripples from different directions interact and cause wave conditions that are very difficult to model mathematically. Various simplified theories and wave spectral models of ocean waves have been introduced in the literature, such as Airy waves (linear waves), higher-order Stokes waves, and irregular waves represented by wave spectra. A 3D simulation software can simulate first-order (Airy wave) and second-order (Stokes wave) regular waves in deep and finite-depth water. Moreover, uni- or multi-directional irregular waves can be modeled with the linear superposition approach. Linear waves are considered the most superficial ocean waves and are based on the assumption of a homogeneous, incompressible, and inviscid fluid and irrotational flow. The wave amplitude is also assumed to be small compared with the wavelength and water depth; consequently, the linear free surface condition is applied [22–28].

The wind is another important factor contributing to both wind-induced waves and loads on marine structures when the superstructure (the portion above the mean water surface) is significant. The wind characteristics include the speed, velocity of the mean wind, and the turbulence or gusts expressing the mean rate's time-varying wind speed. Most of the ocean surface energy is due to the wind-generated waves resulting from the wind blowing over a vast expanse of the fluid surface [22].

To investigate both the strength and direction of the ocean kinetics, we measured the physical marine properties, such as waves and ocean currents, from the water surface to a depth of 10 m near the installation point (IP) for a 15-day observation period. Marine physics was investigated at the south sea for setting up the experiment. The flow velocity and direction of seawater were measured at the surface, at mid-depth, and at the bottom of the sea by an acoustic Doppler current profiler (ADCP). From the statistical analysis of the raw data obtained with the ADCP [22], the maximum flow velocity at the surface, mid-depth, and bottom was 15.1, 13.4, and 16.1 cm/s, respectively, and the flux emergence rate under the velocity of 10 cm/s at the surface, mid-depth, and bottom of the sea was 99.1%, 99.3%, and 98.7%, respectively. Also, the emergence rate for the south direction flow at the surface, mid-depth, and bottom was 61.9%, 57.6%, and 56.2%, respectively, while the emergence rate for the north direction flow at the surface, mid-depth, and bottom was 38.2%, 42.4%, and 43.8%, respectively. Therefore, the south direction flow's emergence rate was higher than that of the north direction flow. Based on the ISO tidal analysis, the ocean currents' progressive vector was calculated at the surface, mid-depth, and bottom for a 15-day observation period.

3.2.4 Simulation and Experimental Setup

Numerical simulations of the aerodynamics, structure, and hydrodynamics of the floater and the mooring system were performed using the hydrodynamic simulation package. The simulations were performed with the same experimental model and settings used for the concept design verification presented in a previous paper [12]. Based on the IPMC sample performance results with the conductive graphene-based solution and the marine physics investigation, we optimized the total active area of the ocean kinetic energy-harvesting module with the IPMC material. A portable power system utilizing the ocean currents' vertical waves was designed and fabricated with an electrochemical material consisting of IPMC to supply electricity to stand-alone offshore plants. The ocean kinetic energy-harvesting module consists of nine vertical and nine horizontal components with ten cells. Water movements are a combination of transverse and longitudinal elements. The transverse segment accounts for the vertical motion of waves, and the longitudinal component refers to the horizontal action of the oceanic currents, as shown in Figure 3.3c. Each vertical piece has a sub-buoy that is sufficient to float at the surface of seawater; each component consisting of ten cells has an

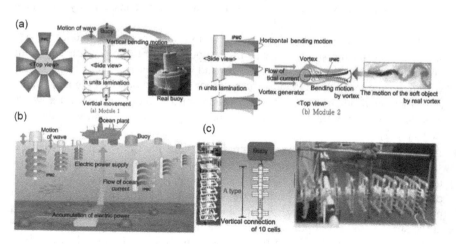

FIGURE 3.3 Conceptual design of the ocean kinetic energy-harvesting structure with an electrochemical material consisting of IPMC. (a) Compact movable power system. (b) The ocean kinetic energy-harvesting structure-based IPMC. (c) A transverse component for the vertical waves, and a longitudinal component for the horizontal ocean currents.

Energy Harvesting System Based on IPMCs

electricity collector connected in parallel and located on the sub-buoy; and each cell has a waterproofed rectifying electricity collector connected in parallel to the four pairs of electrodes that clamp the IPMC sheets. The entire enclosure is waterproof. Both the vertical and horizontal components have a sub-buoy sufficient to float on the seawater's surface, and each part composed of ten cells has an electricity collector in parallel and located on the sub-buoy. The main buoy, which comprises 18 components, consists of an electrical collector in parallel with a power measurement module and a communication module with a photovoltaic power supply that wirelessly sends data to a monitoring system.

Figure 3.4 shows a photograph of the ocean kinetic energy-harvesting system comprising a main buoy and 18 modules installed in the ocean and modeled with a 3D simulation software. However, all geometries of the system simulation were created in a 3D drawing software. The part design and assembly modules in the software were used to establish a three-dimensional parameter model with physical and geometric characteristics. Hereafter, the solid model of the IPMC system created in the drawing environment is imported directly into the 3D simulation software to calculate the dynamic movements. Each IPMC system module is symmetric concerning a vertical plane passing through its center and enclosing a water body with a free surface.

The main buoy was assumed to be stationary; meanwhile, the modules could arbitrarily move according to the wave surface and be connected to the cables' main buoy. Besides, IPMCs were attached to each module and changed shape when the modules were displaced. The movements of the modules and IPMCs under wave attacks are incredibly complicated and are evaluated with response amplitude operators (RAOs). The RAOs show a module's behavior in wave conditions, which is calculated by solving a set of linearized equations based on the Morison equation in different directions. These equations describe the three types of displacement motions (surge, sway, and heave) and three rotational motions (roll, pitch, and yaw), as shown in Figure 3.5.

Consequently, each module's mathematical model was considered floating under the influence of the linearized hydrodynamic fluid wave loading,

 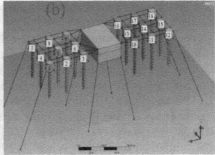

FIGURE 3.4 The ocean kinetic energy-harvesting system including a main buoy and 18 modules (a) installed in the ocean and (b) modeled in 3D simulation software.

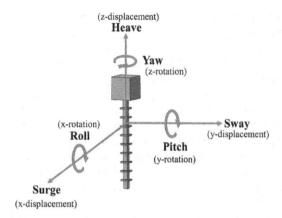

FIGURE 3.5 The six degrees of freedom of the displacement modules at the free surface of the wave.

as described in detail in reference [22]. The hydrodynamic forces are composed of radiation forces and wave excitation forces. The radiation fluid loading causes body motions and is calculated by investigating the radiated wave field from body motions. The incident wave acting on the body is assumed to be harmonic and of small amplitude compared to its length. Also, the fluid is considered ideal, incompressible, and irrotational. Hence, the potential flow theory is used. It is also assumed that the movement and deformation of the IPMCs are directly proportional to each module's direction, which means that the more the modules move, the more electricity the IPMC system generates. The experimental results obtained in both the library and ocean also unify that the electricity performance output of the IPMC strongly depends on the module's movement. Therefore, this research aims to model the displacement of the modules to identify the most suitable assembly method for the system to improve the module's activity, which would enhance the performance of the electricity harvesting module. Moreover, the system's structural stability under wave attacks was also calculated to avoid destruction in typhoons or resonant frequencies.

The numerical simulation of the IPMC system was accomplished by performing frequency domain analyses with the 3D simulation software. In this research, a series of numerical simulations were conducted to investigate the effect of wave forces and moments on the system at various wave heights, wave periods, and wave directions, based on the measured experimental data as mentioned above. In all simulations, deep-sea conditions were considered with different frequencies ranging from 0.03 to 0.93 Hz at 5 m submergence depth. This submergence depth was selected for the simulation since it is similar to the oceanic depth when the IMPC system was assembled in the experiment. Meanwhile, by default, the controlled frequency range in the 3D simulation software has been described in detail in reference [22]; this equally spaced the specified number of frequencies between a minimum value based on the water depth and maximum mesh size. Alternatively, the manual entry can be defined, in which case the wave frequencies can be applied to either a single- or multi-structure selection. The frequency range from 0.03 to 0.93 Hz also covered the frequency range measured from the experiment's assembled IPMC system. For these reasons, the frequency range from 0.03 to 0.93 Hz was selected for investigation in this research. The 3D diffraction calculations considering the interactions between the modules and waves were performed. The responses to the movement (added mass, damping, and motion operators) and to the various wave fields (incoming, diffracted, and radiated waves) are influenced by the seabed, i.e., by the water depth and by the geometry of the seabed [23–28]. A second floating or fixed structure would also influence the ship response and the wave fields. When the sloping bottom is included in the panel model as a second structure resting on the seabed, it will influence the waves and the ship response. This influence, however, may be different from the actual effect of a sloping seabed. The wave conditions in the 3D diffraction analysis and the modules' resistance (added mass and damping coefficients) are described in the following sections.

3.3 RESULTS AND DISCUSSION

3.3.1 The Pressure and Motion Results

In fluid mechanics, a displacement occurs when an object is submerged in a fluid; the displaced fluid volume can be determined by integrating over its submerged surface. The hydrostatic force can be calculated by combining the hydrostatic pressure over the body's wetted surface up to the still water level. The hydrostatic moments are taken about the center of gravity of the body. When dealing with the frequency domain problems, we are concerned with small-amplitude motions about the equilibrium floating position. Thus, the body's wetted surface becomes time independent and the hydrostatic forces and moments about the mean position of the body must be computed. The forces and moments acting on the body directly cause the motion of the body in the fluid. In the 3D simulation software, the simulated pressure and activity obtained by solving the hydrodynamic equations can be visualized and displayed. Pressure and movement may be added. As mentioned above, both the experimental and simulation analysis results showed that the modules' activities were directly proportional to the amplitude of the incidental wave. As a result, this manuscript only shows the simulation results corresponding to the wave amplitude of 0.3 (m). Figure 3.6 plots the pressure and the IPMC system's motion under wave attacks; the corresponding value of the incident wave amplitude is 0.3 m at a frequency of 0.48 Hz. Waves attack all of the modules and the main buoy. The main buoy is assumed to be fixed; meanwhile, the modules are free to move on the wave surface. The results present the modules' movement due to incoming waves, proportionating to the amplitude from −0.413 to 0.580 m.

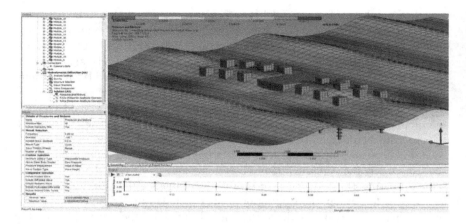

FIGURE 3.6 Water pressures and wave motions around the IPMC system installed in the ocean.

3.3.2 Effect of Wave Direction and Frequency on the Movement of the Modules

The IPMC system includes 18 energy-harvesting modules; however, this chapter only evaluates the motion of module 1, module 5, and module 9 due to the system's symmetry. The direction of the incident wave varies from −180° to 180° (the direction of the wave is determined by the angle created between the wave vector direction and the x-axis). The RAOs are illustrated in Figures 3.7–3.10, including the displacements and rotations of module 1, module 5, and module 9 around the x-, y-, and z-axes, which are, respectively, named surge, sway, heave, roll, pitch, and yaw as shown in Figure 3.4. Each module's amplitude fluctuates dramatically for the frequencies, especially in the frequency range between 0.5 and 0.8 Hz. The incident waves are responsible for the strong oscillations of the modules along the z-axis (heave); meanwhile, the rotations around the z-axis (yaw) are not significant at any given frequency and direction. The modules' surge motion decreases as the frequency increases in almost all wave directions; however, this trend is not evident at a frequency of 90° because this direction is perpendicular to the x-axis, resulting in a strong sway and a weak surge motion as described in Figure 3.9. The modules' roll and pitch motions also oscillate with different amplitudes when the incoming wave's frequency or direction varies. However, this amplitude's maximum value is small—approximately 8°—due to the modules' symmetric shape. Consequently, we should focus on increasing the heave movement to enhance the modules' energy-harvesting performance. The large-amplitude activity of a body can be observed if resonance occurs [27]. Many previous studies on WECs investigated the maximum power of a particular device by using resonance phenomenon to achieve high power absorption levels. However, as illustrated in Figures 3.7–3.11, the heave movements are powerful, corresponding to a frequency range between 0.5 and 0.8 Hz, and the maximum

Energy Harvesting System Based on IPMCs 45

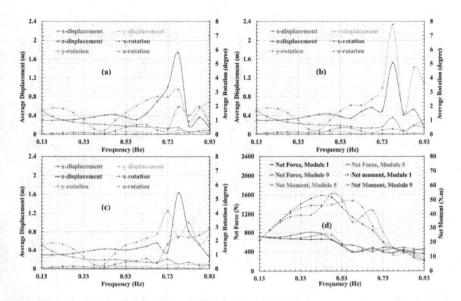

FIGURE 3.7 Response amplitude operators of (a) module 1, (b) module 5, (c) module 9, and (d) net forces/moments at the wave direction of zero degrees.

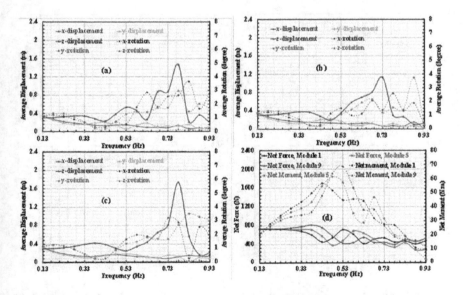

FIGURE 3.8 Response amplitude operators of (a) module 1, (b) module 5, (c) module 9, and (d) net forces/moments at the wave direction of 45°.

46 Ionic Polymer–Metal Composites

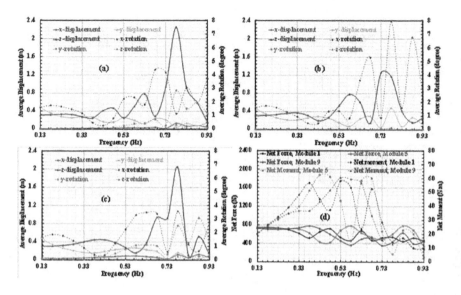

FIGURE 3.9 Response amplitude operators of (a) module 1, (b) module 5, (c) module 9, and (d) net forces/moments at the wave direction of 90°.

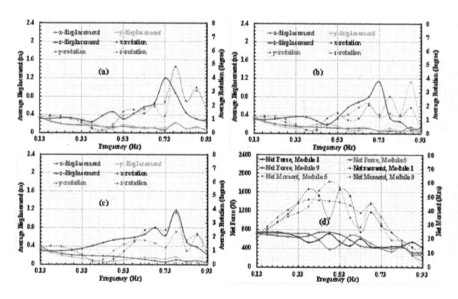

FIGURE 3.10 Response amplitude operators of (a) module 1, (b) module 5, (c) module 9, and (d) net forces/moments at the wave direction of 135°.

Energy Harvesting System Based on IPMCs

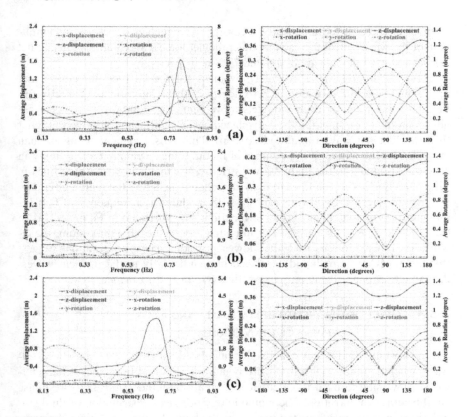

FIGURE 3.11 The average displacement and rotation of 18 modules with a variety of frequency ranges and directions, corresponding to (a) 2.13 m^2, (b) 2.33 m^2, and (c) 2.43 m^2 of wetted surface area for each module.

amplitude of the modules is approximately eight times the amplitude of the incident wave. These movements can cause an instability in the IPMC system, eventually destroying the IPMC system. As a result, we should consider designing the energy-harvesting system's structure to improve its performance and guarantee its stability.

The 3D simulation software computes the hydrodynamic fluid wave loading on a floating or fixed rigid body using three-dimensional radiation and diffraction theory. The hydrodynamic forces are composed of radiation forces and wave excitation forces. The radiation fluid loading is due to body motion and may be calculated by investigating the radiated wave field from body motions. The active or wave excitation loading that induces movement is composed of diffraction forces due to the incident wave's scattering and the Froude–Krylov force due to the pressure field in the undisturbed incident wave. The net forces and moments of the flow acting on module 1, module 5, and module 9, obtained by integrating the pressure over the wetted surfaces, are illustrated in Figures 3.7d, 3.8d, 3.9d, and 3.10d. The modules also oscillate dramatically according to frequency, direction, and amplitude of the waves attacking each module; the net force's maximum value and net moment on each module are 800 N and 70 Nm, respectively.

Figure 3.11a shows the average movements of the 18 modules concerning the frequency at the direction of zero degrees. These movements generally follow the same trend as each module's movement: The rotation movements are not remarkable; meanwhile, the z-displacement is significant and always higher than the incident wave's amplitude (0.3 m). The x-displacement is higher than the y-displacement due to the number of module lines arranging accordingly in each direction and to the main buoy restricting the incident wave (three lines in the x-direction and seven lines in the y-directions, including the main buoy). The average displacement and rotation of the 18 modules at a frequency of 0.48 Hz fluctuate approximately as harmonic functions, as illustrated in Figure 3.11a. At 0°, −180°, and 180°, the surge, yaw, and pitch motions reach a maximum; meanwhile, the other emotions are at their lowest positions. These observations suggest that the IPMC system should be placed perpendicularly to the incident waves' direction to enhance the energy-harvesting system's performance.

3.3.3 Effect of the Wetted Surface Area and Mass on the Movement of the Modules

The hydrostatic fluid forces can also be calculated using the 3D simulation software; when combined with the hydrodynamic forces and body mass characteristics, they may be used to calculate the small-amplitude rigid body response equilibrium mean position. The methodology utilizes a distribution of fluid singularities over the mean wetted surface of the body. Figure 3.11 shows the comparison of the average displacement and rotation of the 18 modules with several frequency ranges and directions, corresponding to (1) 2.13 m^2, (2) 2.33 m^2, and (3) 2.43 m^2 of wetted surface area for each module. The wetted surface is a part of the body in contact with the fluid when it is in equilibrium or steady state. For stability, the body's mass must equal the amount of water displaced, and the center of gravity and buoyancy must be along the same vertical line as no external constraining forces are being applied. This means that each module's mass relates directly to the wetted surface; thus, this affects the module's movement due to the diffraction and radiation wave forces acting on the module, as mentioned in previous reports [23–28]. Generally, the simulated results show that the modules' displacement is proportional to the wetted surface area at most frequencies and directions, as shown in Figure 3.11. However, the trend is opposite near the resonance frequency because, for a body of given volume, the smaller wetted surface area corresponds to the lighter mass. When resonance occurs, the diffraction and radiation wave forces acting on the module increase dramatically; therefore, the module can escape the wave surface. As a result, a lighter module has a higher amplitude of movement than a heavier module. The modules' rotation is minimal: The average amplitude of the 18 modules in the roll, pitch, and yaw motions is only approximately 1° at a frequency of 0.48 Hz, as represented in Figure 3.11.

3.3.4 PERFORMANCE TEST OF THE OCEAN KINETIC ENERGY-HARVESTING MODULES

Based on the simulation results, the IPMC system was assembled to take full advantage of the harvested energy and guarantee stability against wave loads. Consequently, the plan was set perpendicularly to the direction of the incident wave. The main buoy was fixed, and the modules were connected to the main buoy and each other with cables. IPMC materials were attached around the modules' axes to quickly fluctuate when the modules move along the z-direction (the vertical direction) to take full advantage of the modules' high movement in this direction presented in the simulation results. The assembly of IPMCs in the module and the IPMC system are shown in Figure 3.4. Each module includes ten layers of IPMC materials arranged in a vertical line. The total surface area of the IPMC materials of the 18 modules is $4.32\,m^2$ and is uniformly distributed in every module, which is all submerged under the wave surface. The ocean kinetic energy-harvesting modules produced electrical energy after installation in the sea on May 20, 2019. The recorded data were measured and saved at 10-minute intervals over 296 days.

Figure 3.10a shows the extremely variable data obtained on May 20 and July 30, 2019. At a given the time of day, the average power density oscillated above $180\,\mu W/m^2$ and peaked at $280\,\mu W/m^2$ at 18:00 on July 30, 2019. However, each day's average power density was stable, approximately $245\,\mu W/m^2$ during the 296 days, as shown in Figure 3.10b. Compared with other mechanical WECs that convert the energy in ocean waves into electrical power, the IPMC system seems to produce less energy; nevertheless, the IPMC material is an innovative electroactive polymer that makes electrical energy from a bending motion in a simple cantilever configuration. Also, the IPMC is an inexpensive and durable material; therefore, the harvested energy's power can be advantageously improved by increasing the IPMC active area.

3.4 CONCLUSIONS

In this study, the world's first movable ocean kinetic energy-harvesting module consisting of IPMC material was developed and applied to converting the ocean's kinetic energy into electricity. The IPMC system was simulated using the CFD method to evaluate its movement and structural stability and to identify the optimal setup to enhance its energy-harvesting performance. Tests under real sea conditions were also performed to analyze the electrical harvesting performance of the IPMC system.

In general, the modules' displacement was proportional to the wetted surface area at most frequencies and directions. However, the trend was opposite near the resonance frequency due to the large increase in wave forces. The modules' rotation was minimal: The average amplitude of the 18 modules in the roll, pitch, and yaw motions was only approximately 1° at a frequency of $0.48\,Hz$. Also, the z-displacement was significantly higher than the amplitude of the incident wave

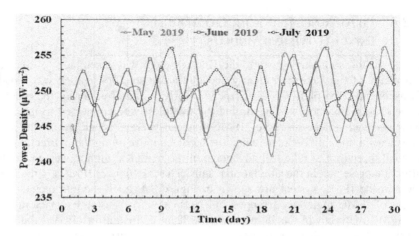

FIGURE 3.12 The power density of the energy-harvesting system for 3 months.

and the displacements in other directions. Based on the simulation results, the experimental IPMC system was set perpendicular to the incident wave's direction to take full advantage of the harvested energy. The total surface area of the IPMC materials submerged under the wave surface of the 18 modules is 4.32 m^2 and is uniformly distributed in every module. The ocean kinetic energy-harvesting modules that produce electrical energy was installed in the sea on May 20, 2019. The experimental results showed that the average power density at a given time of the day intensely oscillated above 180 μW/m^2 and reached a peak value of 280 μW/m^2 at 18:00 on July 30, 2019. Also, the data recorded for 296 days (from May 20, 2019, to March 17, 2020) verified that the average power density for each day was stable, approximately 245 μW/m^2, and the degradation of the IPMCs' electrical performance in the long term of operation is trivial (see Figure 3.12). However, it is further recommended that the growth of algae and barnacles on the modules should also be diminished to maintain high efficiency in generating electricity; therefore, every module should be covered by a protective layer to prevent the penetration of algae and barnacles for maintaining the IPMCs' electrical performance.

In conclusion, the development of ocean kinetic energy-harvesting systems with a conductive graphene-based solution of IPMC expressed many advantages compared with mechanical WECs for generating electric power from sea kinetic energy as IPMC is an inexpensive, durable material and creates electricity. However, the initial concept's implementation to the commercial manufacturing stage has been challenging, slow, and expensive. It requires more work for the thorough evaluation of abnormal phenomena such as resonance or the disruption of electrical energy harvesting owing to the growth of algae and barnacles on the module, which affected the bending motion of the IPMC. Also, the specific power of the IPMC system is still small compared to other devices; thus, it is necessary to identify solutions to enhance its performance.

Future research will focus on performing the analysis of buoy systems for varying current and wave conditions to test the system design's structural stability in the ocean wave environment in case of typhoon or extreme weather, including unusual, severe, or unseasonal weather. We will also investigate and measure the IPMC material properties to determine their movements and electrical characteristics. The entire process of activities of the complex IPMC system and wave amplitude on its activities will be physically modeled and validated by using the experimental results.

In the next chapter, the reader will find a new method of IPMC soft actuator synthesis using single-walled carbon nanotubes, and also micro-parts handling with IPMC microgripper. In the next chapter, the characterization process of IPMCs is also shown.

REFERENCES

[1] E. Lejerskog, C. Boström, L. Hai, R. Waters, M. Leikon, 2015, "Experimental Results on Power Absorption from a Wave Energy Converter at the Lysekil Wave Energy Research Site". *Renewable Energy*, Vol. 77, pp. 9–14.

[2] S. H. Salter, 1989, "World Progress in Wave Energy-1988". *International Journal of Ambient Energy*, Vol. 10, pp. 3–24.

[3] B. Teillant, R. Costello, J. Weber, J. Ringwood, 2012, "Productivity and Economic Assessment of Wave Energy Projects through Operational Simulations". *Renewable Energy*, Vol. 48, pp. 220–230.

[4] A. F. O. Falcão, 2010, "Wave Energy Utilization: A Review of the Technologies". *Renewable & Sustainable Energy Reviews*, Vol. 14, pp. 899–918.

[5] D. V. Evans, 1981, "Power from Water Waves". *Annual Review of Fluid Mechanics*, Vol. 13, pp. 157–187.

[6] J. Falnes, 2002, "Optimum Control of Oscillation of Wave-Energy Converters". *International Journal of Offshore and Polar Engineering*, Vol. 12, pp. 1–9.

[7] J. Sjolte, G. Tjensvoll, M. Molinas, 2013, "Power Collection from Wave Energy Farms". *Applied Sciences*, Vol. 3, pp. 420–436.

[8] J. Scruggs, P. Jacob, 2009, "Harvesting Ocean Wave Energy". *Science*, Vol. 323, pp. 1176–1178.

[9] M. Rahm, O. Svensson, C. Boström, R. Waters, R. M. Leijon, 2012, "Experimental Results from the Operation of Aggregated Wave Energy Converters". *IET Renewable Power Generation*, Vol. 6, pp. 149–160.

[10] M. J. Khan, G. Bhuyan, M. T. Iqbal, J. E. Quaicoe, 2009, "Hydrokinetic Energy Conversion Systems and Assessment of Horizontal and Vertical Axis Turbines for River and Tidal Applications: A Technology Status Review". *Applied Energy*, Vol. 88, pp. 1823–1835.

[11] G. Bracco, E. Giorcelli, G. Mattiazzo, 2011, "ISWEC: A Gyroscopic Mechanism for Wave Power Exploitation". *Mechanism and Machine Theory*, Vol. 46, pp. 1411–1424.

[12] S. K. Park, J. W. Ahn, J. K. Lee, S. H. Park, H. M. Kim, K. W. Park, G. O. Hwang, M. K. Kim, S. H. Baek, G. S. Byun, 2014, "An Ionic Polymer Metal Composite Based Electrochemical Conversion System in the Ocean". *International Journal of Electrochemical Science*, Vol. 9, pp. 8067–8078.

[13] Ch. Youngsu, K. Hubert, P. Maurizio, 2013, "Energy Harvesting from Underwater Base Excitation of a Piezoelectric Composite Beam". *Smart Materials and Structures*, Vol. 22, pp. 1–14.

[14] J. Kwaśniewski, I. Dominik, F. Kaszuba, 2014, "Energy Harvesting System Based on Ionic Polymer-Metal Composites – Identification of Electrical Parameters". *Polish Journal of Environmental Studies*, Vol. 23, pp. 2339–2343.

[15] N. D. Vinh, M. K. Kim, 2020, "A Study of the Movement, Structural Stability, and Electrical Performance for Harvesting Ocean Kinetic Energy Based on IPMC Material". *Processes*, Vol. 8, p. 641.

[16] N. D. Vinh, M. K. Kim, 2017, "Ocean-Based Electricity Generating System Utilizing the Electrochemical Conversion of Wave Energy by Ionic Polymer-Metal Composites". *Electrochemistry Communications*, Vol. 75, pp. 64–68.

[17] H.İ. Yamaç, A. Koca, 2018. "Numerical Analysis of Wave Energy Converting Systems in Case of Using Piezoelectric Materials for Energy Harvesting". *Journal of Marine Engineering & Technology*, Vol. 1, pp. 138–149.

[18] Y.C. Chow, Y.C. Chang, C.C. Lin, J.H. Chen, S.Y Tzang, 2018. "Experimental Investigations on Wave Energy Capture of Two Bottom-Hinged-Flap WECs Operating in Tandem". *Ocean Engineering*, Vol. 164, pp. 322–331.

[19] X. Zhang, D. Lu, F. Guo, Y. Gao, Y Sun, 2018. "The Maximum Wave Energy Conversion by Two Interconnected Floaters: Effects of Structural Flexibility". *Applied Ocean Research*, Vol. 71, pp. 34–47.

[20] C. Oh, S. Kim, H. Kim, G. Park, J. Kim, J. Ryu, P. Li, S. Lee, K. No, S. Hong, et al 2019. "Effects of Membrane Thickness on the Performance of Ionic Polymer–Metal Composite Actuators". *RSC Advances*, Vol. 9, pp. 14621–14626.

[21] D.J. Li, S. Hong, O Heinonen, 2013, "Polymer Piezoelectric Energy Harvesters for Low Wind Speed". *Applied Physics Letters*, Vol. 104, pp. 1–6.

[22] ANSYS, Inc, 2014, *ANSYS Fluent® 15: AQWA Theory Manual*, ANSYS, Canonsburg, PA.

[23] J. Kim, S. Yun, Z. Ounaies, 2006, "Discovery of Cellulose as a Smart Material". *Macromolecules*, Vol. 39, pp. 4202–4206.

[24] J. Smagorinsky, 1963, "General Calculation Experiments with the Primitive Equations". *Monthly Weather Review*, Vol. 91, pp. 99–164.

[25] A. Giacomello, M. Porfiri, 2011. "Underwater Energy Harvesting from a Heavy Flag Hosting Ionic Polymer Metal Composites". *Journal of Applied Physics*, Vol. 109, pp. 1–10.

[26] L. A. Weinstein, M. R. Cacan, P. M. So, P. K. Wright, 2012, "Vortex Shedding Induced Energy Harvesting from Piezoelectric Materials in Heating". *Smart Materials and Structures*, Vol. 21, pp. 1–10.

[27] D. V. Evans, R. Porter, 2012, "Wave Energy Extraction by Coupled Resonant Absorbers". *Philosophical Transactions of the Royal Society*, Vol. 370, pp. 315–344.

[28] Y. H. Zheng, Y. M. Shen, J. Tang, 2007, "Radiation and Diffraction of Linear Water Waves by an Infinitely Long Submerged Rectangular Structure Parallel to a Vertical Wall". *Ocean Engineering*, Vol. 34, pp. 69–82.

4 Application of Ionic Polymer Metal Composite (IPMC) as Soft Actuators in Robotics and Bio-Mimetics

Ravi Kant Jain
CSIR-Central Mechanical Engineering Research Institute (CMERI)

CONTENTS

4.1 Introduction 54
4.2 Past Literature Survey on the Development of IPMC Actuators, Modeling, Control and Various Robotic and Biomimetic Applications 55
4.3 Development of Single-Walled Carbon Nanotube (SWNT)-Based IPMC Soft Actuators 61
    4.3.1 Material Requirements 61
        4.3.1.1 Materials 61
        4.3.1.2 Reagent Solutions 62
        4.3.1.3 Functionalization of SWNTs 62
        4.3.1.4 Sulfonation of PEES 62
    4.3.2 Fabrication of IPMCs 62
    4.3.3 Characterization 63
        4.3.3.1 Ionic Conductivity 64
        4.3.3.2 WU, IEC and PC 65
        4.3.3.3 FTIR Study 66
        4.3.3.4 Tensile Strength 67
        4.3.3.5 SEM, EDX and TEM Studies 69
        4.3.3.6 Porosity 71
        4.3.3.7 UV–Visible Studies 71
        4.3.3.8 Thermal Analysis 72
        4.3.3.9 Electrochemical Characterization 72

         4.3.3.10 Electromechanical Characterization............................ 74
4.4 Development of IPMC Soft Actuator-Based Robotic System for
    Robotics Assembly 78
    4.4.1 IPMC-Based Microgripper for Remote Center Compliance
          (RCC) Assembly.. 78
    4.4.2 An IPMC-Based Two-Finger Microgripper for Handling
          Millimeter-Scale Components.. 81
    4.4.3 An IPMC-Based Artificial Muscle Finger Actuated
          through EMG.. 82
    4.4.4 Robotic Micro-Assembly Using IPMC Microgrippers 85
4.5 Conclusions.. 87
Acknowledgment 87
References... 87

4.1 INTRODUCTION

In the recent past, the robotics and biomimetics areas have shown growing interest in using electroactive polymer (EAP) soft actuators and sensors. An ionic polymer–metal composite (IPMC) is a class of EAP (ionic polymer-type) soft actuators. This is basically softly driven by a low voltage (0–5 VDC) and can work quickly and provide bending responses [1,2]. The bending response can be utilized in different robotic and biomimetic applications such as robotic assembly, miniature parts assembly, biomimetic mechanics, aerospace, hepatic joints and medical applications [3–6] because they are lightweight, easily processed, flexible, highly sensitive and resilient, biocompatible, etc. They can also be used as artificial muscle-like actuators for various human affinity and biomedical applications [7,9]. To cater such needs, IPMCs are therefore promising for developing/manufacturing IPMC soft actuators for different applications. Principally, IPMCs consist of a thin ionomer composite polymer membrane with metal (Pt or Au) electrodes deposited on both faces. The protons on the anionic groups covalently bonded to the backbone of composite polymer membrane are typically exchanged for metal cations, and the element is soaked in a solvent, usually water [7]. Consequently, the polymeric composite material-based actuators suffer associated problems such as short cycle lifetime and less response time [10,11]. Therefore, it is required to use electrolytes with polymeric membrane so that the performance of soft actuators can be enhanced. In this chapter, the development, characterization and various applications of IPMCs are proposed for robotic and biomimetic applications. These IPMCs provide several advantages in their characteristics such as high proton conductivity, high ion exchange, high water uptake, excellent film forming capacity and large bending response. During the development of the ionic polymer composite, non-volatile materials are used and are characterized by their high ionic conductivities. These will provide the benefit of large actuation performance of IPMC in air. Thus, various combined chemical and electromechanical properties are studied in this chapter, which may improve the bending rate, displacement and repeatability in the developed IPMC membrane.

This chapter is focused on the following points:

a. Development of IPMC soft actuators based on single-walled carbon nanotubes (SWNTs).
b. Chemical, mechanical, thermal and electromechanical characterization of IPMC-based soft actuators.
c. Development of an IPMC soft actuator-based robotic system for robotics assembly.

This chapter is organized as follows: a survey of the past research on the development of IPMC actuators, modeling, control and various robotic and biomimetic applications is presented in Section 4.2. The details for the development of IPMC soft actuators based on single-walled carbon nanotubes (SWNTs) are presented in Section 4.3. In Section 4.4, the development of an IPMC soft actuator-based robotic system for robotic assembly is described. The conclusions are provided in Section 4.5.

4.2 PAST LITERATURE SURVEY ON THE DEVELOPMENT OF IPMC ACTUATORS, MODELING, CONTROL AND VARIOUS ROBOTIC AND BIOMIMETIC APPLICATIONS

In the past, several researchers have attempted to develop various aspects of IPMC actuators, such as actuator development, modeling, control and various robotic and biomimetic applications. Shahinpoor et al. [12] started working on IPMCs where initially a continuum electromechanical theory for the dynamic deformation of ionic polymeric gels in the presence of an imposed electric field is discussed. Using this theory, a theoretical model has been proposed based on the electro-osmosis, electrophoresis and ionic diffusion of various species in gel conditions. Further, Shahinpoor et al. [13] attempted mathematical modeling of IPMCs where a number findings with respect to using IPMCs as biomimetic sensors and actuators are presented. Using these findings, Shahinpoor et al. [14] reported the capabilities of IPMCs as biomimetic sensors and actuators, where the IPMC characteristics have shown the potential of both actuators and sensors. A data acquisition system is used to find the parameters such as voltage, current and displacement in real time. Nemat-Nasser et al. [15] described the properties of IPMCs for use as soft actuators and sensors that could be applicable in robotics, biomimetics, bio-inspired devices and artificial muscles. Laurent et al. [16] designed a swimming microrobot using IPMCs, where an undulatory motion is given by the IPMC. A theoretical model has also been developed for verifying the motion. Further, Shahinpoor et al. [17] presented an experimental study of IPMCs in various cation forms for actuation behavior, where tetrabutylammonium (TBA) and tetramethylammonium (TMA) organic compounds have mixed Li^+ for coating on Nafion films. Madden et al. [18] presented the related work of artificial muscles technology, where the understanding, principal and advantages of locomotion behavior for biomimetic fish and insects are discussed. Bhat et al. [19] attempted theoretical modeling based on precision position control of an IPMC strip in a cantilever configuration, where a closed-loop control method was applied for reducing the overshoot during actuation. Bar-Cohen [20] gave emphasis on the potential of IPMC actuators as artificial muscles, where various characteristics including lightweight and large actuation behavior (stretching, contracting or bending) are identified and, by changing shape and size, the microrobots can be constructed. Kim et al. [21] explored the possibility of the use of hydrogels as artificial muscles, where PAAc/PVSA co-polymer hydrogels have shown electrically induced contractions with associated deformation that can be used in the development of artificial muscle actuators. Kim et al. [22] developed a wireless undulatory tadpole robot using IPMC actuators, where a biomimetic undulatory motion of the tail is demonstrated. Lee et al. [23] reported on the design and numerical analysis to find the free strain and blocked stress of IPMC-based linear actuators, which are important parameters of muscle-like actuators. Further, Lee et al. [24] attempted to design and fabricate an IPMC actuator for a flapping device where an IPMC-based flapping mechanism is developed and the flapping behavior has also been tested. Chung et al. [25] fabricated an IPMC actuator by applying coating of silver nano-powders on Nafion films, which provided a good adhesion between the metal electrodes and polymer membrane without surface roughening pretreatment conditions. Malone et al. [26] developed IPMC actuators which are tailored in unusual geometries and integrated with other freeform fabricated active components. Kim et al. [27] carried out a work on the analytical model of a segmented IPMC propulsor, where the model on the dynamics of the flexible IPMC bending actuator has been formulated for accommodating the relaxation behavior of IPMCs. Fang et al. [28] developed IPMC actuators, and the dynamic behavior of these actuators was analyzed for better actuating performance which has shown the second-order polynomial behavior. The open-loop and closed-loop controllers were also designed for controlling active catheter systems. Konyo et al. [29] discussed various kinds of microrobotic systems developed using IPMC actuators and sensors, such as hepatic interface for virtual tactile displays, distributed actuation devices and a three-DOF soft micromanipulation device. Pugal [30] attempted to fabricate a self-oscillating IPMC bending actuator, where the theoretical model was developed using time-dependent differential equations and the simulated data were verified with the experimental data. Ji et al. [31] attempted to synthesize Nafion-117 membranes with polypyrrole/alumina composite to analyze and verify the equivalent cantilever beam and equivalent bimorph beam models. Chen et al. [32] developed a control-oriented and physics-based model for IPMC actuators, where an infinite-dimensional transfer function relating the bending displacement to the applied voltage is derived where an H_∞ controller is designed based on the reduced model and applied to tracking control.

Chen et al. [33] presented a non-linear control-oriented model for IPMC actuators, where the proposed model is designed using the non-linear capacitance of the IPMC and a non-linear partial differential equation (PDE) is applied, which

IPMCs as Soft Actuators

is mapped with steady-state voltage conditions. Further, Chen [34] submitted his PhD thesis on IPMC artificial muscles and sensors, where the control system perspective of IPMCs was discussed. The generated bending, twisting and cupping motions are applied in fish robotics. Chen et al. [35] presented a bio-inspired robotic manta ray behavior using an IPMC where the IPMC acts as artificial muscles to mimic the swimming behavior of the manta ray. Chen et al. [36] also presented a review on IPMC artificial muscles in bio-inspired engineering applications where the underwater propulsion applications were discussed. Jung et al. [37] attempted to fabricate an IPMC-based nano-composite actuator, where fullerene-reinforced nanoparticles were coated on a Nafion film. The harmonic responses led toward stability during actuation. Saher et al. [38] applied O_2 plasma treatment method for depositing a thin platinum electrode layer on the membrane surface during electroless chemical metal plating process. The performance of actuation, displacement, force and operational life of the plasma-treated IPMNC actuator were compared with other IPMC actuators. Bahramzadeh et al. [39] presented the application of IPMCs to the measurement of sensitive curvature and the dynamic structures where the effect of voltage recovery of IPMCs on sensing signal was studied by applying the deformation in the form of ramp functions at very slow rates of curvature variation. Mukherjee et al. [40] carried out the dynamics analysis of IPMC actuators, where the simulation results showed the potential for the use of IPMC flapping wings as a viable contestant toward developing the insect-scale flapping wing of micro-air vehicles. Gutta [41] presented the modeling and control of an IPMC actuator, where a local coordinate system approach is used for identifying the bending of each segment of IPMC. A large deflection model approach is combined with clumped RC model to find the dynamics of the IPMC. Panwar et al. [42] attempted to study the dynamic mechanical, electrical and actuation properties of IPMC actuators using polyvinylidene fluoride (PVDF)/polyvinylpyrrolidone (PVP)/polystyrene sulfonic acid (PSSA) blend membranes. Price [43] studied the properties of composite materials having conductive polymers as a constituent, where the multilayer electroactive polymer actuators consisting of polypyrrole films electro-polymerized on a passive polymer membrane core were exploited as actuators. Lantada et al. [44] obtained a non-linear response by developing a model using artificial neural networks (ANNs), which helps in the encapsulation of IPMCs during electromechanical characterization of transducers. Lee et al. [45] addressed an approach for controlling IPMCs using a wireless power link between the IPMC and a remote unit using microstrip patch antennas toward transmitting the power on electrode surface of the IPMC. A frequency modulation of the microwave is also proposed for selecting the actuation portion of the IPMC where the matching patch antenna pattern is located.

He et al. [46] prepared an organic–inorganic hybrid Nafion/SiO_2 membranes-based IPMC with various tetraethyl orthosilicate (TEOS) contents, which can be used for actuating artificial eyes. Kang et al. [47] presented an adaptive feedforward control architecture for IPMC toward deflection control, where adaptive identification was executed by changing the dynamic behavior over time and input voltage using a gradient descent method. Shi et al. [48] focused on an electromechanical model of the IPMC and examined the deformation and actuating force of the IPMC equivalent to a cantilever beam. A concept of underwater legged microrobot was also demonstrated. Leang et al. [49] focused on the control approaches for precision control of IPMC actuators. An inverse model is developed to obtain the feedforward inputs for precisely tracking the desired output trajectory/path. Caponetto et al. [50] proposed an enhanced fractional-order transfer function model for IPMC actuators, where the IPMC model was analyzed by conducting experiments. Luca et al. [51] focused on the fabrication and modeling of smart devices for the realization of post-silicon applications, where various kinds of modeling methods were discussed. Sasaki et al. [52] focused on the self-sensing control of Nafion-based IPMC actuators, where the self-sensing controllers were designed using feedforward, feedback and two-DOF techniques. These help in enhancing the performance of bending curvatures. Okazaki et al. [53] developed soft actuators using IPMCs driven with ionic liquids, where it is estimated that the bending ability continues for a long time in spite of the atmospheric humidity condition. Chen [54] developed a mathematical model of the IPMC actuator for a stable second-order dynamical system where the hysteresis is controlled and the position error is minimized. Kwaśniewski et al. [55] focused on the energy harvested by the IPMC, which can be used for powering small electronic devices. This has been done with the help of series and parallel combinations of different IPMC samples. Farid et al. [56] attempted the forward kinematic modeling and simulation of IPMC actuators for bionic knee joint, where the forward kinematic model was developed for the manipulator by applying the geometric coordinate system method and the simulations were carried out using Wolfram Mathematica software. Dominik et al. [57] attempted a physics-based geometrically scalable model to develop a control system where the relation between the actuating voltage and the tip displacement was discussed with help of a transfer function. Mutlu et al. [58] focused on a fully compliant micro-stage with built-in actuation, which has been fabricated as one piece inspired from twining structures in nature, and employed soft robotic modeling and finite element modeling approaches to find the mechanical output. Tang et al. [59] attempted to fabricate low-cost and high-performance IPMC actuators, where sulfonated polyphenylsulfone (SPPSU) membranes were mixed with different degrees of sulfonation (DS) for increasing their ion exchange capacity, water uptake and ion conductivity. Palmre et al. [60] attempted to develop an IPMC actuator with nanostructured electrodes to achieve highly enhanced electromechanical performance over existing flat-surfaced electrodes, where the formation and growth of the nanothorn assemblies at the electrode interface led to a dramatic improvement in both actuation process and blocking force at low driving voltages. Hirano et al. [61] examined the influence of the hydration level on the electromechanical behavior of Nafion-based IPMCs, where by conducting experiments, the bending force, electrical pulses and relative humidity were measured. Simaite [62] developed a hybrid membrane using poly(3,4-ethylenedioxythiophene)/polystyrene sulfonate (PEDOT:PSS)-based actuators, where the drop casting fabrication method was applied. Sun et al. [63] attempted to integrate the static and dynamic models of the IPMC actuators. The transfer function for the IPMC mechanical response to an electrical signal was obtained. This model could be helpful in low-frequency, power-efficient applications. Carrico et al. [64] applied a three-dimensional (3D) fused filament additive manufacturing technique for developing soft active 3D structures, which were developed through layer-by-layer method. The performance of 3D-printed IPMC actuators was compared with Nafion-based IPMC actuators. Nam [65] attempted to manufacture IPMC actuators with blended Nafion membranes, where the electromechanical performance was compared with Nafion membrane-based IPMCs in terms of thermal and mechanical properties. Shen et al. [66] reported an IPMC actuator having multiple-shape memory effect, which can perform complex motions by two external inputs. This is called a soft multiple-shape memory polymer–metal composite (MSMPMC) actuator having multiple DOFs. Hong et al. [67] examined the electrochemical and morphological characteristics of IPMCs by varying the morphology of their metal composite component such as conductive network composite. The dependence of electrochemical properties on CNC nanostructure and the mechanoelectrical performance of IPMC sensors as a function of CNC morphology were also demonstrated. Peterson et al. [68] demonstrated IPMC-based energy harvesting from coherent fluid structures, where the bending deformation of a cantilevered IPMC strip followed axisymmetric bending of an IPMC annulus due to the impact of a vortex ring. Haq et al. [69] presented a comprehensive review of IPMCs, where their fundamentals, fabrication processes, characterization and applications were discussed. Kazem et al. [70] proposed a non-linear neural network black box model (NBBM) for predicting the bending motion of IPMCs, where the applied electrical voltage characteristics and the working conditions were taken into account. Bian et al. [71] focused on doping with $BaTiO_3$ nanoparticles for fabricating high-performance IPMC actuators, where the performance of IPMCs was evaluated in terms of their static mechanical properties, water uptake, surface resistivity, electrochemical impedance and actuation behavior. Chen [72] presented a review work on IPMC-enabled underwater robots, where the design, modeling and fabrication perspectives were discussed. A physics-based and control-oriented model of IPMC actuator was also presented. Pasquale et al. [73] attempted to fabricate IPMCs and ionic polymer–polymer composites (IP2Cs) using Nafion®-117, where the electrode coating method was adopted with different metals (Ag, Au or Pt). The effects of electrodes on the mechanical, thermal and electromechanical behaviors of IPMCs and IP2Cs were also discussed. Carrico et al. [74] presented a comprehensive review of some smart polymeric and gel actuators (such as IPMCs) for mechatronics and robotics applications, where the fundamentals, fabrication processes and motion control were discussed. Porfiri et al. [75] attempted a mathematical modeling for finding the counterion size that has to be present in metal particles during the actuation of IPMCs so that the steric effect of the back-relaxation of IPMCs can be reduced by decreasing the counterions concentration near the electrodes. Boldini et al. [76] established a mathematical model for estimating the onset of the pull-in instability during electrostatic actuation to enhance the experimental IPMC performance. Bernat et al. [77] presented an approach for tuning multiple models of an online identifier by integral mapping to IPMC where the extension of the estimation law of additional mapping between parameters and measurable signals is applied where significantly improvement has been done in transient responses of IPMC without increasing feedback gain. Sainag et al. [78] developed an optimal position control of IPMC using particle swarm optimization method, where a comparative study of PI (proportional–integral) tuning parameters based on conventional Ziegler–Nichols (ZN) and PSO tuning technique was carried out, which showed the improvement of step response characteristics such as rise time, settling time and maximum overshoot. Wang et al. [79] attempted developing a molding process or integrating rod-shaped IPMC actuators with the help of a soft silicone rubber structure. Further, the IPMC-embedded tube designs for the development of robotic-assisted manipulation were also demonstrated. Zhao et al. [80] proposed a design of biomimetic flapping air vehicle by combining the IPMC with the bionic beetle flapper, where the flapping displacement under different voltage conditions and frequency was experimented. Shahinpoor et al. [81] showed the potential of the looped IPMC haptic feedback sensor that can be enabled using advanced technology in robotic surgery. An IPMC feedback sensor can be used for normal grasping and manipulation of bodily organs and tissues. These looped IPMC haptic force feedback sensors can also be applied as a smart skin for the development of human-like dexterous and soft manipulation. Khan et al. [82] fabricated a silver nano-powder embedded with Kraton polymer actuator, where the chemical, electrochemical and electromechanical characterization was also carried out. Further, Khan et al. [83] attempted to manufacture sulfonated graphene oxide and sulfonated poly(1,4-phenylene ether-ether-sulfone) blended with polyvinylidene fluoride-based IPMCs with enhanced performance, and the characterization was carried out. This was carried out using the film-casting method. The considerable potential as a soft actuator material for robotic and biomimetic applications is also shown. Biswal et al. [84] fabricated a multilayered structure of IPMC soft actuators by using the chemical decomposition method, where an increase in the number of layers on the base material (Nafion-117) increased the actuation capability of the IPMC actuator compared with the single-layer IPMC actuator. Further, Biswal et al. [85] developed a non-linear mathematical model for the analysis of the dynamic behavior of IPMC cantilever actuators, where the large bending deformations occurred under AC excitation voltages. Tabatabaie et al. [86] attempted to achieve the undulating linear actuation and sensing by slit IPMCs, and their combination with twisting and bending created the 3D soft robotic manipulation and undulation motion like an elephant trunk. Further, Tabatabaie [87] attempted to design an artificial soft robot in his PhD thesis, where similar motions of an elephant trunk were demonstrated. The soft undulating robot is developed using a combination of several IPMC sections. Further, Tabatabaie [88] also tried different configurations of IPMCs, which are used in various applications as slit, cylindrical, tubular and helical IPMC actuators and sensors, and haptic and tactile feedback sensors. Zhang et al. [89] discussed the

IPMCs as Soft Actuators

main characteristics and consideration of IPMCs for design and implementation in robotic artificial muscles and biomimetic robots that offer perception of next-generation muscle-powered robots. Gupta et al. [90] focused on a kinematic and dynamic model of a biomimetic robot actuated by a soft IPMC actuator. The relation between the charge stored and the applied voltage to actuate the IPMC was derived by the ordinary differential equation for the RC circuit.

Further, Wang et al. [91] attempted to develop a simplified static model for studying the sensing properties of IPMCs with various sizes, where the effect of dimensions on the Pd–Au-type IPMC membranes with various thicknesses, widths and lengths and deformation sensing performances were tested. MohdIsa et al. [92] proposed different sensing methods, which are divided into three different categories such as active sensing, passive sensing and self-sensing actuation (SSA). The active sensing methods can be used for measuring only one of IPMC-generated voltage, charge or current, whereas the passive methods can be used for measuring variations in IPMC impedances or can be utilized in capacitive sensor element circuits. The SSA methods can be applied simultaneously for sensing and actuation on the same IPMC. Annabestani et al. [93] presented a comprehensive review on the applications of ionic electroactive polymer (i-EAP) soft actuators as active elements of microfluidic micropumps, microvalves and micromixers, where the related works presented by various researchers in this field were composed. Tripathi et al. [94] developed IPMC-based artificial muscles by using the physical vapor deposition fabrication technique. Li et al. [95] performed characterization and actuation of IPMCs by varying the thickness and length of the membrane, and the actuation methods are applied in three different conditions—no treatment, treatment in deionized water and treatment in a NaCl solution. Gonçalves et al. [96] performed an electrochemical analysis of Nafion-based IPMCs operated in dry and wet environments, where the mass transport is strongly affected by both the ionomeric channel width and the interaction between the species and the Nafion side chains. Yang et al. [97] put an emphasis on the prediction of the actuation property of Cu-based IPMCs using back-propagation neural networks, where the displacement and blocking forces were analyzed. Ma et al. [98] demonstrated the high-performance behavior of IPMCs for mimicking biological motions such as petal opening, closing, tendril coiling, uncoiling and high-frequency wing flapping, and these behaviors can be used for biomedical devices and bio-inspired robotics.

4.3 DEVELOPMENT OF SINGLE-WALLED CARBON NANOTUBE (SWNT)-BASED IPMC SOFT ACTUATORS

4.3.1 MATERIAL REQUIREMENTS [99]

4.3.1.1 Materials

Poly(1,4-phenylene ether-ether-sulfone) pellets (Tg 192°C); 1-ethyl-3-methylimidazolium tetrachloroaluminate, ≥95% (IL); N,N-dimethylacetamide, ≥99% (NMA) (Sigma-Aldrich, the USA); carbon nanotubes, single-walled, min.

90%, OD: 1–2 nm, length: 5–30 μm (SRL Pvt. Ltd., India), tetraammineplatinum(II) chloride monohydrate [Pt(NH$_3$)$_4$Cl$_2$·H$_2$O (crystalline)] (Alfa Aesar, USA); ammonium hydroxide (25%); hydrochloric acid (67%); sodium nitrate and sodium sulfate (Merck Specialities Pvt. Ltd., India); and borohydride (NaBH$_4$) (Loba Pvt. Ltd., India) were taken.

4.3.1.2 Reagent Solutions

The aqueous solutions of HCl (2 M), NaNO$_3$ (1 M), tetraammineplatinum(II) chloride monohydrate (0.04 M), NH$_4$OH (5.0%) and NaBH$_4$ (5.0%) were prepared using demineralized water (DMW).

4.3.1.3 Functionalization of SWNTs

The acid form of commercially available carbon nanotubes was obtained by sonicating in 1:3 (v/v) concentrated nitric acid and sulfuric acid for 4 hours at 45°C. The resultant solid was washed thoroughly several times with DMW until the pH value reaches 6–7. Now, the acid form SWNT-COOH was sonicated in aqueous NaOH (5 mM) for 2 minutes to convert it into sodium salts (SWNT-COO$^-$Na$^+$). After washing, the black solid of SWNT-COO$^-$Na$^+$ (SWNT) was collected and dried at 50°C. Finally, the functionalized SWNT-COO$^-$Na$^+$ (250 mg) was added to 10 mL NMA and the mixture was mechanically agitated followed by sonicating for degassing to make a homogeneous dispersion. The schematic diagram of functionalization of carbon nanotubes for the development of soft actuator is shown in Figure 4.1.

4.3.1.4 Sulfonation of PEES

Sulfonation of base polymer PEES was carried out, where a 5 g PEES was dissolved completely in excess of concentrated H$_2$SO$_4$ (125 mL) by strong mechanical agitation (450 RPM) for a time interval of 36 hours. In this development, the concentrated H$_2$SO$_4$ was used as both a solvent and a sulfonating agent. After constant stirring up to the determined reaction time to obtain white sulfonated PEES (SPEES) strings, the polymer solution was gradually precipitated in ice-cold DMW. The decanted SPEES strings were repeatedly washed several times with DMW until the pH of the wash water was approximately 6–7. The SPEES is hydrophilic in nature and shows a considerable swelling during the neutralization process. This swollen sulfonated PEES strings were dried under the oven thermostat for over 24 hours at 60°C and then at room temperature.

4.3.2 Fabrication of IPMCs

To fabricate the SPEES polymer membrane by the conventional solution casting method, 16 wt% of the polymer solution was prepared in NMA at 45°C under continuous stirring for obtaining a viscous solution. A homogeneous casting blend was prepared by the addition of 200 μL of IL and an appropriate amount of functionalized SWNT dispersion (0.8 g in 25 mL NMA) to the SPEES solution under constant stirring for 24 hours. On ensuring complete blend dissolution, the solution was degassed by sonication for 30 minutes before casting. Then,

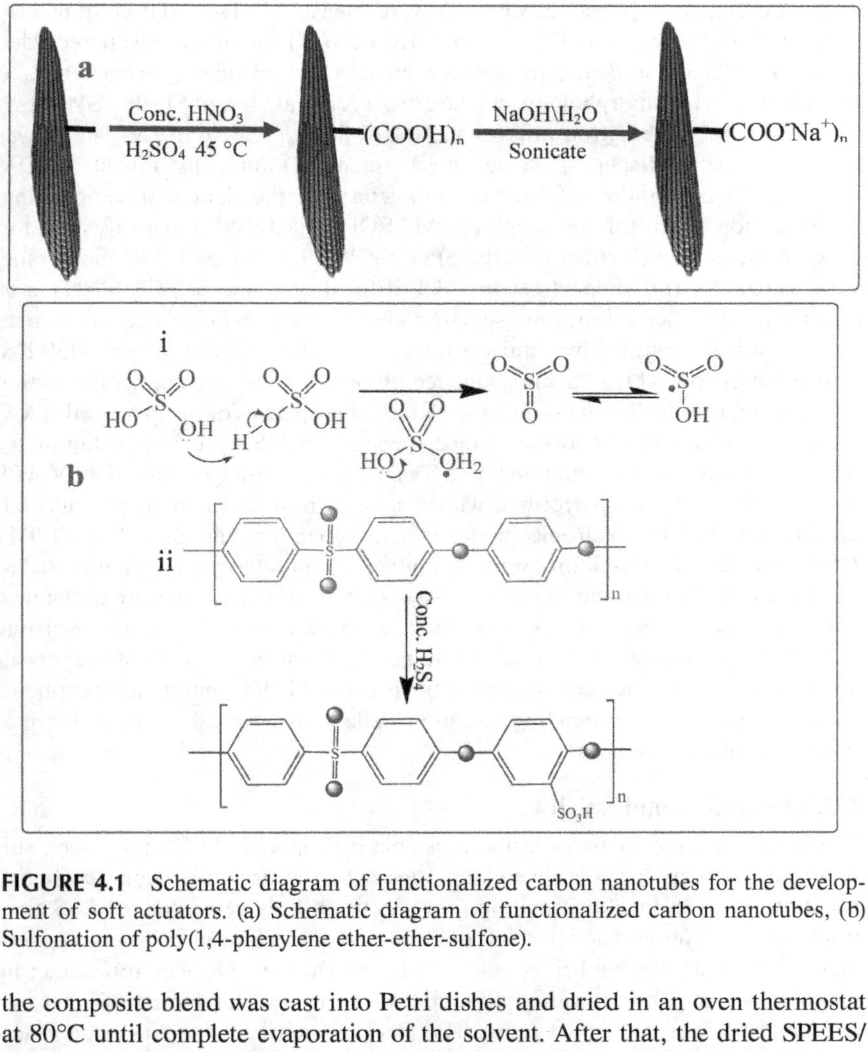

FIGURE 4.1 Schematic diagram of functionalized carbon nanotubes for the development of soft actuators. (a) Schematic diagram of functionalized carbon nanotubes, (b) Sulfonation of poly(1,4-phenylene ether-ether-sulfone).

the composite blend was cast into Petri dishes and dried in an oven thermostat at 80°C until complete evaporation of the solvent. After that, the dried SPEES/IL/SWNT (SPEES-SWNT) membrane was peeled off and washed with DMW and acetone several times before electroless plating of Pt metal.

To fabricate the IPMC actuator, the SPEES-SWNT-based polymer membrane was sandwiched between conductive metal ions (Pt metal) as an electrode by the electroless plating or chemical reduction method as reported by Khan et al. [100].

4.3.3 Characterization

The chemical and electromechanical properties, surface morphology and chemical composition of the SPEES-SWNT-Pt composite-based IPMCs were studied by a variety of techniques. The water uptake (WU) and proton conductivity (PC) of the developed polymer membranes were measured. The Fourier transform infrared (FTIR) spectra of SPEES and SPEES-SWNT membranes were recorded between 500 and 4000/cm using a spectrometer (PerkinElmer Spectrometer). To observe the surface morphology and the cross-sectional view of SPEES-SWNT-Pt IPMC, a scanning electron microscope (SEM JOEL, JSM-6510LV, Japan) was used. An energy-dispersive X-ray spectroscope (Oxford Instruments INCA, X.act, S. No. 56756, the UK) was used to determine the elemental composition. Transmission electron microscopy (TEM JEOL, JEM-2100, Japan) was used to observe the detailed structure of the SPEES-SWNT-Pt-based IPMC membrane. UV–visible absorption spectroscopy of PEES, SPEES and SPEES-SWNT was employed using the PerkinElmer spectrophotometer Lambda 25. The mechanical stability was determined by a universal testing machine (UTM; model: H50 KS, Shimadzu Corp.), with a 25 mm gauge length between the grips under the testing speed of 5 mm/min. To investigate the electrical properties of the proposed IPMC membrane, electrochemical impedance spectroscopy (EIS), cyclic voltammetry (CV) and linear sweep voltammetry (LSV) at triangle voltage input of ±2 V with a step of 50 mV/s were performed with the Autolab 302N modular potentiostat/galvanostat. Before electromechanical characterization, the SPEES-SWNT-Pt IPMC membranes (1×3 cm^2) were immersed in an aqueous solution of 0.2 M NaOH at room temperature for 6 hours so that the countercations were exchanged with Na$^+$. The successive steps of bending response at 0–4.5 V DC and overtime at 4.5 V were analyzed using a laser displacement sensor. To find out the repeatability and to check the performance of the proposed IPMC, multiple experiments were conducted, such as bending actuation, deflection hysteresis, normal distribution and load characterization.

4.3.3.1 Ionic Conductivity

To obtain ionic conductivity in thermostable hydrophobic PEES polymers, sulfonic acid groups (–SO$_3$H) were introduced to the aromatic, non-fluorinated backbone by a sulfonation reaction. The electrophile is generated, and an electrophilic substitution reaction takes place at one of the four positions of the aromatic ring between ether bridges and SPEES repeat unit. The electron-attracting nature of the neighboring sulfonyl group lowers the electron density of other two aromatic rings in the repeat units. The electron-rich benzene ring then reacts with the electrophile to give the sulfonated product. The DS (%) of pure SPEES, SPEES-SWNT and SPEES-SWNT-Pt IPMC was found to be 126, 114 and 112, respectively, and was high enough to enable sufficiently high proton conduction. Noticeable changes in the DS (%) of different composite membranes were known to occur with the introduction of the SWNTs and Pt metal in the polymer. It was observed that with an increase in DS (%), the SPEES polymer exhibited excessive swelling as well as partial dissolution in water. To increase the mechanical stability, functionalized SWNTs were added to the SPEES solution during the fabrication of ionic polymer composite as discussed in the experimental section. To check the stability, the SPEES-SWNT ionomeric membrane was immersed in hot water (80°C) for 24 hours, and it was found that the membrane was completely

IPMCs as Soft Actuators

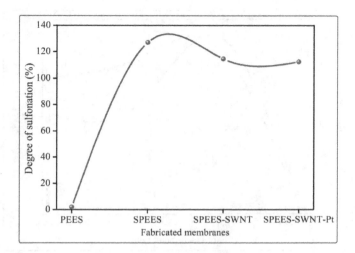

FIGURE 4.2 Ionic conductivity of the developed soft actuator [99].

stable for further characterization. This observation indicated that the SPEES-SWNT with higher DS (%) was suitable for membrane-based IPMC actuators and their applications (see Figure 4.2).

4.3.3.2 WU, IEC and PC

The percentage of WU of polymer membranes is a critical property having a good effect to improve the PC of the ionic membrane and dielectric constant of the IPMC membrane. The PC increases with the increase in the WU of polymer membranes, and at the same time, the mechanical stability of the membrane declines. An elevated WU is vital for the actuation of an IPMC membrane, because of the movement of the hydrated cations along with the water molecules, when a low electric potential is applied to the IPMC. Figure 4.3 shows that the maximum WU for SPEES, SPEES-SWNT and SPEES-SWNT-Pt polymeric membranes was found to be 162%, 92% and 80%, respectively, after immersion in DMW for 24 hours. The WU of the pure SPEES is much more than that of the blended SPEES-SWNT and SPEES-SWNT-Pt ionomeric membranes due to the high IEC. Compared to the pure SPEES membrane, the SPEES-SWNT and SPEES-SWNT-Pt membranes exhibited a lower WU. The addition of SWNTs decreases the porosity and the number of acidic sites, and hence, the capacity of the membrane to retain the water decreases with the increase in the mechanical stability. Notably, the WU of final SPEES-SWNT-Pt IPMC membrane was much higher than that of Nafion-based IPMCs. It is very important for improving the performance of IPMC actuators.

A higher IEC of the polymer composite materials means a larger amount of water molecules move toward the negative electrode by the action of hydrated cations; electro-osmosis and electrophoresis cause a greater volume expansion toward the cathode side, resulting in the consequent larger bending deformation.

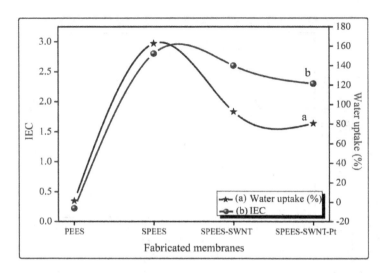

FIGURE 4.3 Water uptake, proton conductivity and ion exchange properties [99].

The high IEC enables that Pt particles can more easily be embedded deep within the porous surface by the chemical reduction plating method and reduce the resistance, increase the electric current and increase the capacitance of an IPMC membrane actuator. The IEC values of SPEES, SPEES-SWNT and SPEES-SWNT-Pt were found to be 2.8, 2.6 and 2.5 meq/g of dry membrane, respectively. The SPEES-SWNT-Pt-based IPMC membrane has a higher IEC than Nafion-based IPMC membranes. This is because the SPEES-SWNT-Pt membrane actuator has a larger number of sulfonic acid groups. The data obtained from the analysis of IEC and WU of the fabricated SPEES-SWNT-Pt polymer-based IPMC actuator revealed that the proposed actuator has a higher water and ionic content compared to Nafion-based IPMC actuators. The higher IEC and WU of the SPEES-SWNT-Pt polymer-based IPMC actuator confirm the better performance relative to other conventional polymer-based IPMC actuators. The maximum PC of the SPEES-SWNT-Pt-based IPMC was found to be 2.321×10^{-2} S/cm. The high PC of the SPEES-SWNT-Pt membrane enables the better performance because cations move quickly toward the cathode, to create an imbalance in the pressure inside the membrane. Thus, the rate of actuation of the IPMC membrane increases with the high PC, which leads to the quick movement of more hydrated cations toward the cathode side, showing a large displacement and fast actuation. The electric current of an IPMC increases with the increase in the PC of the ion exchange polymer membrane. It has been found that the high capacitance and electric current increase the actuation under applied AC and DC voltages.

4.3.3.3 FTIR Study

A FTIR spectroscopy analysis was performed on the SPEES and SPEES-SWNT polymer membranes to confirm the successful introduction of sulfonic

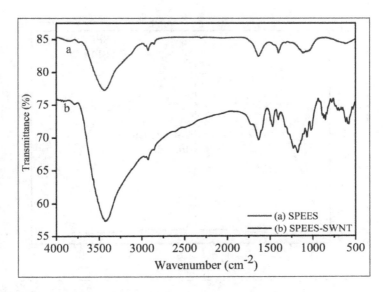

FIGURE 4.4 FTIR spectroscopy analysis [99].

acid groups to the PEES and the functionalization of carbon nanotubes. The absorption band at 3434 cm^{-1} is due to the O–H stretch of –SO$_3$H and the absorbed moisture, as shown in Figure 4.4. The characteristic absorption peaks of –SO$_3$H appear at 1089 and 1034 cm^{-1}, indicating the existence of O=S=O and S=O stretches, respectively. The absorption bands at 1210 and 1580 cm^{-1} could be assigned to the C–O stretch of the acid and the C=C stretch of the SWNT backbones, respectively. The absorption peak at 1720 cm^{-1} confirms the existence of the carboxylic acid groups. The absorption bands at 1574 and 1378 cm^{-1} suggest the formation of SWNT-COO$^-$Na$^+$. The characteristic absorption bands of carboxylate salt may also be overlapped with the C=C stretch of the SWNT backbones [99].

4.3.3.4 Tensile Strength

The mechanical properties of the fabricated polymer composite membrane such as Young's modulus and tensile strength play a significant role in the actuation performance of the IPMC actuator. Both the SPEES and SPEES-SWNT-Pt membranes of width 20 mm and thickness 0.15 mm were fixed into a UTM with a fixed gauge length of 25 mm. As the membranes elongated with loads, the displacement of the two end points was continuously recorded to attain the mechanical strength. Figure 4.5 shows the stress–strain curve of the SPEES and SPEES-SWNT composite membranes and their mechanical properties. Young's modulus and ultimate tensile strength of the SPEES and SPEES-SWNT composite membranes were calculated by the stress–strain curve, as given in Table 4.1 [99]. To find out the value of Young's modulus, the slope of the stress–strain curve was used. Ten slope values at different points

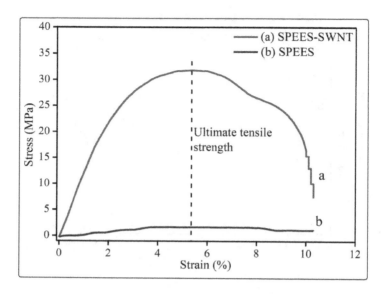

FIGURE 4.5 Tensile property of SPEES and SPEES-SWNT composite membranes [99].

TABLE 4.1
Mechanical Properties of SPEES and SPEES-SWNT Ionomeric Polymer Membranes [99]

Membrane	Young's Modulus (MPa)	Ultimate Tensile Strength (MPa)	Elongation at Break (%)
SPEES-SWNT	932.6	31.8	10.1
SPEES	88.77	1.8	10.3

of the stress–strain curve were taken from the starting point to the maximum stress value. Subsequently, the average of these ten slope values was used to calculate Young's modulus of the fabricated ionomeric membranes. The ultimate tensile strength was calculated at the maximum stress point. Young's modulus and ultimate tensile strength of the pure SPEES membrane were 88.77 and 1.8 MPa, respectively, whereas the values were 932.62 and 31.8 MPa, respectively, for the SPEES-SWNT composite membrane. The results demonstrate that the composite blend of the SPEES-SWNT membrane provides high molecular rigidity and mechanical strength, which leads to a higher tensile modulus relative to the SPEES membrane. The SPEES-SWNT shows improved mechanical properties than the pristine SPEES polymer membrane because SWNTs have excellent mechanical strength as well as superior electrical properties [99].

IPMCs as Soft Actuators 69

4.3.3.5 SEM, EDX and TEM Studies

Figure 4.6 shows the surface and cross-sectional morphologies of the SPEES-SWNT-Pt composite polymer-based IPMC membranes. On the surface of SPEES-SWNT membrane, the Pt particles were homogeneously distributed and covered the whole interface boundaries of the IPMC membrane. The Pt particles as an electrode are densely deposited on the surface of the SPEES-SWNT IPMC actuator. The Pt diffusion layers can enhance the interfacial adhesion between the electrode and polymer membrane in this IPMC actuator. Some microcracks are clearly shown on the SPEES-SWNT-Pt IPMC membrane surface as a result of the cracking of the Pt electrode layers during the sample drying before SEM analysis. The cross-sectional micrographs show the two layers

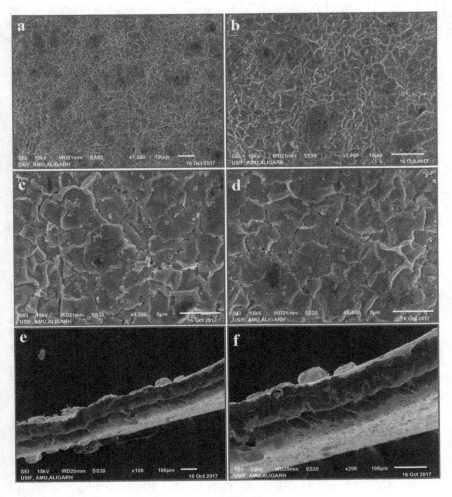

FIGURE 4.6 SEM analysis of SPEES-SWNT IPMC actuator [99].

of Pt electrodes deposited on both the sides of the SPEES-SWNT composite membrane [99].

Figure 4.7 shows the results of the EDX analysis. The EDX spectrum of the SPEES-SWNT-Pt membrane surface reveals the characteristic peaks of elements and the composition of carbon, oxygen, sulfur, chlorine and platinum on the surface of the fabricated IPMC actuator. The characteristic peaks of the Na element provide the functionalization of SWNTs into SWNT-COO$^-$Na$^+$. The large amount of the Pt on the surface confirms the excellent and uniform coating of the metal electrode on the surface of the SPEES-SWNT-Pt-based IPMC membrane. This also suggests that the performance of the proposed actuator is better. The TEM micrographs of the SPEES-SWNT-Pt-based IPMC membranes are shown in Figure 4.7c and d. The black spots show SPEES on a gray background of alkyl-modified SWNTs, and Pt as electrode is homogeneously distributed and covers the gray background of SPEES and SWNT matrices of the SPEES-SWNT-Pt IPMC membrane. The TEM image reveals the existence

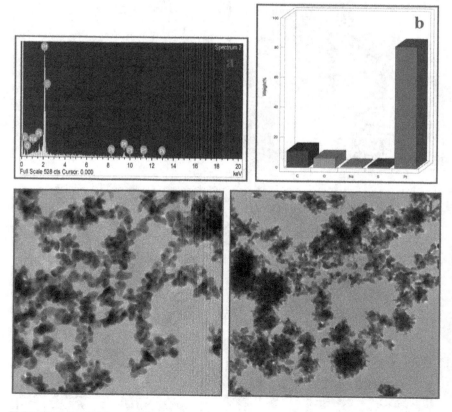

FIGURE 4.7 EDX and TEM analysis of SPEES-SWNT-Pt-based IPMC membranes [99].

IPMCs as Soft Actuators

and homogeneous composite blend of the SPEES, SWNT and Pt electrode in the fabricated IPMC membrane.

4.3.3.6 Porosity

The BET analysis was used to study the porous nature of the SPEES membrane as shown in Figure 4.8. The ionic polymer SPEES membrane exhibited the typical N_2 adsorption–desorption curve for porous materials. The total pore volume and surface area of the SPEES ionic polymer membrane were found to be 6.477×10^{-3} cm^3/g and 16.970 m^2/g, respectively. In addition, the BET results confirm the porous structure of the SPEES membrane, which facilitates ion transport within the ionic polymer membrane, thus leading to a high-performance IPMC actuator [99].

4.3.3.7 UV–Visible Studies

The UV–visible absorbance patterns of PEES, SPEES and SPEES-SWNT determine the chemical interaction between the SPEES and SWNT composite in the developed IPMC membrane, as shown in Figure 4.9. This shows that the absorbance peaks at 274, 278 and 287 nm may be attributed to the characteristic absorbance peaks of PEES. The additional absorbance peaks at 290 and 294 nm appeared after sulfonating the PEES polymer. This was also used as a base ion exchange polymer for the fabrication of IPMC actuators. The absorbance pattern of SPEES-SWNT-Pt-based IPMC shows an additional absorbance peak at 282 nm with the diminishing of other peaks. This confirms the chemical interaction between the SPEES and SWNTs in the IPMC membrane. The additional absorbance peaks with the disappearance/minimization or overlapping of the

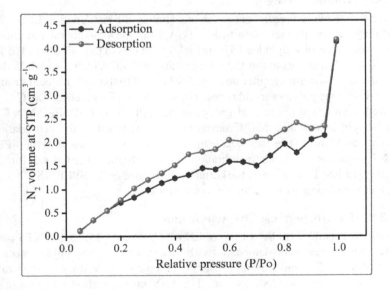

FIGURE 4.8 Identification of porosity [99].

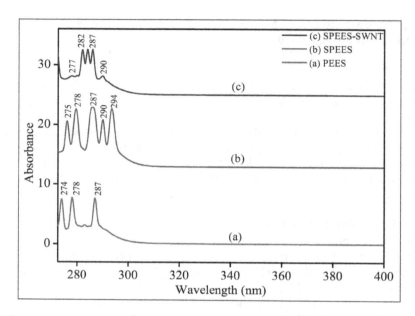

FIGURE 4.9 UV studies of PEES, SPEES and SPEES-SWNT membranes [99].

other peaks after the addition of SWNTs and Pt give evidence of the chemical interaction between SPEES, SWNTs and Pt-based composite used for the fabrication of IPMC actuators [99].

4.3.3.8 Thermal Analysis

Figure 4.10 shows the results of the thermal studies of SPEES and SPEES-SWNT ionic polymer membranes examined by TGA. Both the ionic polymer membranes have three stages of weight loss. The initial weight loss in the SPEES and SPEES-SWNT membranes occurs at the temperature of 50°C–200°C, which is due to the removal of water molecules absorbed or bonded to the sulfonic acid groups in the ion exchange polymer membranes. The second stage located at 300°C–450°C resulted from the sulfonic acid groups in the SPEES and SPEES-SWNT. The third weight loss around 500°C onward can be assigned to the degradation of polymer backbones. The thermogravimetric analysis confirms that the SPEES-SWNT composite polymer membrane exhibits a higher thermal stability than pristine SPEES. This suggests the better performance of the SPEES-SWNT ionic polymer membrane even at a higher temperature.

4.3.3.9 Electrochemical Characterization

The electrochemical studies of the SPEES-SWNT and SPEES-SWNT-Pt-based IPMC membranes were demonstrated by the current–voltage hysteresis curve obtained using CV and LSV at a triangle voltage of ±2 V under a scan rate of 50 mV/s in a three-electrode system. The LSV curve is shown in Figure 4.11. This reveals that the SPEES-SWNT-Pt IPMC shows a higher value of current

IPMCs as Soft Actuators

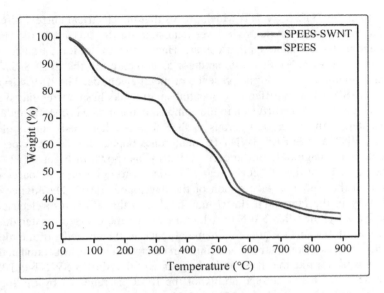

FIGURE 4.10 Thermal studies of SPEES and SPEES-SWNT ionic polymer membranes [99].

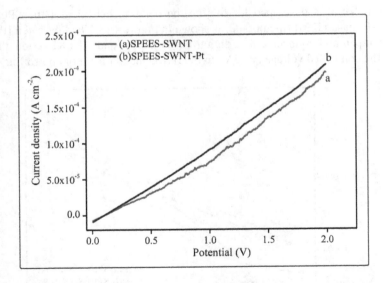

FIGURE 4.11 Current–voltage (I–V) curve of SPEES-SWNT and SPEES-SWNT-Pt-based IPMC membranes [99].

density than the SPEES-SWNT, which proves that the electrical current directly depends on the types of electrode layer deposited on the surface of the IPMC membrane under the same applied voltage. Hence, the rate of ionic transfer was higher in SPEES-SWNT-Pt IPMC actuator as compared to the SPEES-SWNT composite membrane due to the presence of metal electrode. The LSV curve of the SPEES-SWNT with different sequential small peaks has an irregular shape. This indicates that ionic diffusion in the composite membrane is chaotic and may be attributed to the rough and porous surface. The I–V hysteresis curves of the SPEES-SWNT and SPEES-SWNT-Pt membranes appear similar in shape, but different in the magnitude of the current density, as shown in Figure 4.12. The symmetric shape of the CV curves can be assigned to the excellent charge distribution in the whole surface region of the developed IPMC. The higher current density in the fabricated IPMC may be due to the addition of electrically conductive alkyl-modified SWNTs, which increases the charge transfer due to the increase in the surface area for chemical reaction. The benefits of the higher IEC, PC, increased conductivity and uniform electroding of Pt was indicated by the more capacitor like the shape of the I–E curve of the SPEES-SWNT-Pt IPMC membrane. The energy storage ability of the IPMC is reflected by the higher current density, which is responsible for the better actuation performance by the larger deformation of the polymer membrane under applied voltage.

4.3.3.10 Electromechanical Characterization

A testing setup is developed for finding the bending behavior of the SPEES-SWNT-Pt-based IPMC membrane, as shown in Figure 4.13. The fabricated IPMC was clamped and connected to a digital power supply and NI-PXI system. This sends the controlled voltage (±4.5 V) to the IPMC membrane. For controlling this

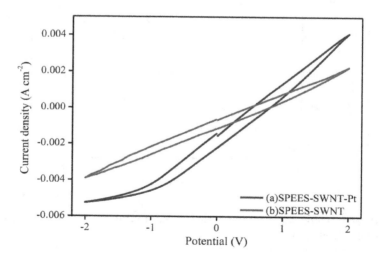

FIGURE 4.12 I–V hysteresis curves of SPEES-SWNT and SPEES-SWNT-Pt membranes [99].

IPMCs as Soft Actuators

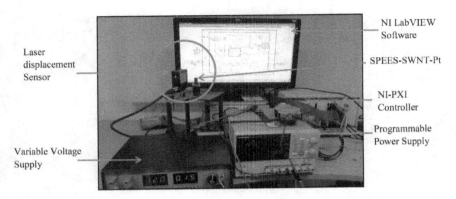

FIGURE 4.13 Experimental testing setup for finding the bending behavior of SPEES-SWNT-Pt-based IPMC membrane.

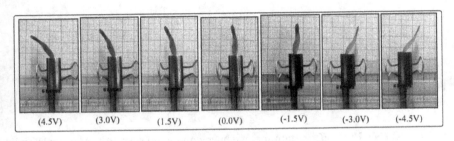

FIGURE 4.14 Bending response of SPEES-SWNT-Pt IPMC actuator [99].

voltage, a VI was designed in the LabVIEW software, where the proportional–integral–derivative (PID) control system feature enabled the incorporation of the PID parameters. An algorithm was also developed for the proper actuation of IPMC. A laser displacement sensor was used for finding the tip displacement of the IPMC membrane. This sensor was activated at 12 V DC and placed in front of the fabricated IPMC. This was interfaced with the NI-PXI system through an RS-485 to the RS-232 converter. The laser displacement sensor continuously checks the displacement position and provides information regarding reaching the desired displacement point.

These experimental deflection data were acquired into the NI-PXI through COM port and NI-VISA interface software module in LabVIEW VI. Simultaneously, the data acquisition (DAQ) card assistant of the NI-PXI system confirms the actuating voltage from the programmable power supply. The bending responses of the SPEES-SWNT-Pt membrane actuator are shown in Figure 4.14. The maximum deflection of the SPEES-SWNT-Pt IPMC was up to 16 mm at 4.5 V DC. Multiple experiments were conducted to investigate the hysteresis behavior, and the average value of deflections at different values of voltage was plotted, as shown in Figure 4.15. To check the performance of the

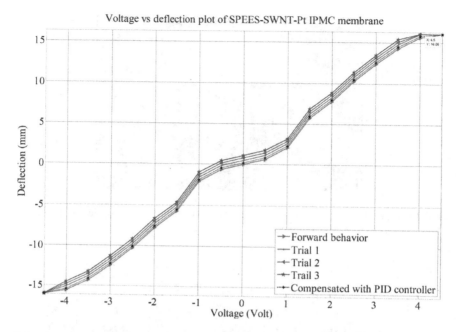

FIGURE 4.15 Experimental performance of bending behavior of SPEES-SWNT-Pt IPMC actuator [99].

SPEES-SWNT-Pt IPMC membrane, we have conducted several experiments and the average deflection values are given in Table 4.2. After multiple experiments for determining the deflection behavior by applying voltage, it was found that the IPMC did not provide the same behavior, which may be due to the partial evaporation of inner solvent by electrolysis under applied voltage or natural evaporation. This may be the main reason behind the hysteresis occurred in the fabricated SPEES-SWNT-Pt-based IPMC actuator. To reduce this hysteresis (error), a PID control system was incorporated in the control system. By enabling this feature and setting up the PID gain parameters, the hysteresis was reduced. For this purpose, an algorithm was also developed, where the controller bandwidth was set by tuning the frequency. When the frequency in the controller was increased, the time duration of the actuation of the SPEES-SWNT-Pt-based IPMC decreased, but at the same time, the stability of the IPMC also decreased. In order to achieve the fast response of IPMC, by proper tuning of gains in the PID controller, the frequency was set. This reveals the steady position between the response time and stability. In this way, the hysteresis was reduced by up to 80%.

A digital weighing/load cell (model: Citizen CX-220; make: India) was taken for load characterization of IPMC membrane, as shown in Figure 4.16. The tip of the SPEES-SWNT-Pt IPMC touches the pan of the load cell while applying voltage through the NI-PXI system, as shown in Figure 4.16. The maximum load-carrying capability of the SPEES-SWNT-Pt-based IPMC membrane was up to

TABLE 4.2
Deflection Data of SPEES-SWNT-Pt IPMC Membrane at Different Applied Voltages [99]

Deflection (mm)	Voltage									
	0 V	0.5 V	1.0 V	1.5 V	2.0 V	2.5 V	3.0 V	3.5 V	4.0 V	4.5 V
d1	0	0.6	2.5	6.0	8.0	10.5	12.6	14.9	15.5	16.0
d2	0	0.7	2.6	6.0	8.2	10.2	12.4	14.2	16.0	16.3
d3	0	0.5	1.9	5.8	8.5	10.6	12.6	14.0	15.4	16.2
d4	0	0.8	2.0	5.5	7.8	10.1	12.7	13.8	15.7	16.1
d5	0	0.7	2.1	5.8	7.6	10.9	12.1	14.5	15.8	15.8
d6	0	0.9	1.8	5.4	7.4	10.0	12.9	14.3	15.4	15.9
d7	0	0.4	1.9	5.8	7.6	10.3	12.4	14.7	15.9	16.0
d8	0	0.6	1.8	5.8	7.5	10.0	12.2	14.6	15.2	16.1
d9	0	0.7	2.2	5.5	7.3	10.2	12.0	14.0	15.4	16.2
d10	0	0.5	2.1	5.8	7.8	10.0	12.0	13.9	15.8	16.0

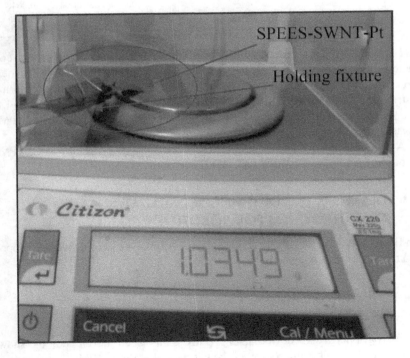

FIGURE 4.16 Testing setup for load characterization.

TABLE 4.3
Experimental Force Data of SPEES-SWNT-Pt IPMC Membrane at Different Voltages

Force (mN)	0V	1V	1.5V	2.0V	2.5V	3.0V	3.5V	4.0V	4.5V
F1	0	0.1185	0.4086	0.8496	2.8263	4.4923	7.3960	9.3903	10.14202
F2	0	0.1029	0.3204	0.9290	2.5950	3.9082	6.4160	9.8205	10.33116
F3	0	0.2097	0.4076	0.6585	2.6185	4.6618	8.3731	8.5495	10.31548
F4	0	0.1960	0.3077	0.8898	1.9482	4.7598	7.4921	9.2365	10.13712
F5	0	0.1607	0.2361	0.4557	2.4186	3.5711	6.3455	9.8039	10.31058
F6	0	0.1313	0.2293	0.5929	2.9449	3.9053	7.3950	10.0401	10.01756
F7	0	0.1577	0.2077	0.5399	2.3823	4.5785	7.6332	9.8196	10.40858
F8	0	0.1401	0.2381	0.8791	2.3490	4.2796	6.7169	9.6647	10.03128
F9	0	0.1313	0.1960	0.7105	2.3863	4.9450	7.5303	9.2757	10.42818
F10	0	0.1411	0.2067	0.8251	2.4225	3.7867	6.7561	9.5687	10.21258

Operating voltage	4.5 V
Mean	10.2334 mN
Standard deviation	0.0648
Repeatability	90.24%

1.03 g at 4.5 VDC. In order to check the repeatability, several trials were conducted and the force data were noted for ten times, as given in Table 4.3. The mean force value was calculated and the normal distribution was found and the normal distribution for the SPEES-SWNT-Pt IPMC was plotted, as shown in Figure 4.17. The narrow shape of the normal distribution suggests that the fabricated IPMC has less error and great repeatability of the forced behavior at 4.5 V DC.

A comparison of different properties of the SPEES-SWNT-Pt-based IPMC membrane actuator with other reported IPMC actuators is given in Table 4.4.

4.4 DEVELOPMENT OF IPMC SOFT ACTUATOR-BASED ROBOTIC SYSTEM FOR ROBOTICS ASSEMBLY

An IPMC soft actuator-based robotic system for robotic assembly was developed, and the description is given below.

4.4.1 IPMC-Based Microgripper for Remote Center Compliance (RCC) Assembly

A microgripper RCC device using IPMCs has been developed, as shown in Figure 4.18. The microgripper constructed by three IPMC fingers (size: 40 mm × 10 mm × 0.2 mm) can actuate in dry environments. These ionic strips were custom-made, and each ionic strip weighed 0.2 g in the ideal condition. These strips were attached to a Perspex sheet by packing tape. The platinum sides

IPMCs as Soft Actuators

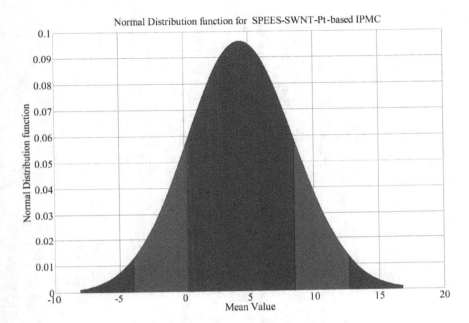

FIGURE 4.17 Normal distribution for the SPEES-SWNT-Pt IPMC [99].

of the strip were connected with wires by copper tape. These wires were connected to a signal amplifier driven with PCI software through a compatible PC. In the compatible PC, output DAC was inserted for converting the digital data into analog data. A trans-amplifier with a power supply was used to amplify the signal output. The computer code generated various waveforms such as sinusoidal, square, triangular and saw tooth signals at desired frequencies and amplitudes of up to 10 V. One end of top thick ionic strip (40 mm × 10 mm × 2 mm) was attached to the RCC device, and the other end was held in a holding device. The input voltage was given to the top thick ionic strip separately. The three fingers were activated by the DAC for holding the thermocol object along with peg, and subsequently, the top ionic strip was separately operated for lifting the object. During this operation, a dial gauge was placed for measuring the peg insertion depth. The depth of peg insertion allows for misalignment during the operation that occurs. For the measurement of misalignment/orientation angle, suitable slip gauges were placed at the bottom.

The RCC device developed using IPMCs facilitates the insertion of peg-in-hole for micro-assembly operations. The effect of compliance during "peg-in-hole" orientation was analyzed using ionic polymer strips. The micromanipulation task was also performed using an IPMC-based microrobotic arm. This shows the potential of RCC using IPMCs. By conducting experiments, the kind of assembly helps in the compensation of misalignment error during robotic assembly.

TABLE 4.4
Comparison Table of SPEES-SWNT-Pt-Based IPMC Membrane Actuator with Other Developed Actuators [99]

Parameter	SPEES-SWNT-Pt	Sulfonated Polyetherimide [101]	Sulfonated Polyvinyl Alcohol [102]	CNT-Based Actuator [103]	Nafion [104]	Kraton-Based IPMC [82]
IEC (meq/g)	2.60	0.55	1.20	0.71	0.98	2.00
PC (S/cm)	2.32×10^{-2}	1.40×10^{-3}	1.60×10^{-3}	5.70×10^{-3}	9.00×10^{-3}	1.30×10^{-3}
WU (%)	80.0	26.4	82.3	25.1	16.7	233.0
Current density (A/cm^2)	4.1×10^{-3}	5.0×10^{-4}	5.5×10^{-3}	–	3.0×10^{-3}	2.5×10^{-3}
Tip displacement (mm)	16.0	2.7	18.5	20.0	12.0	17.0

IPMCs as Soft Actuators

FIGURE 4.18 IPMC microgripper for remote center compliance (RCC) assembly [105].

4.4.2 An IPMC-Based Two-Finger Microgripper for Handling Millimeter-Scale Components

Herein, a two-finger microgripper made of IPMCs has been developed. IPMCs have shown a great potential as high-displacement and lightweight actuators. The low mass force generation capability is utilized for microgripping in a micro-assembly. The IPMC responds to a low voltage in the range of 0–3 V. The material contains an electrolyte that transports ions in response to an external electric field. The IPMC actuation for microgripping is given by applying controlled voltages, where an external electric field generated by a suitable RC circuit causes this deflection. It is found that an IPMC actuates from 1 to 5 seconds. The maximum jaw opening and closing positions of a microgripper are found to be 5 and 0.5 mm, respectively, while IPMCs are in operation. An experimental prototype was developed as shown in Figure 4.19.

The major advantage of a compliant gripping system along with control system is that the mimicking of fingers is done through electrical actuation instead of conventional motor. Each finger can be actuated individually so that a dexterous handling is possible, which allows for precise end effector positioning. When fingers are in operation, the maximum displacement is in the range of 5 mm. The fingers can hold an object having a weight of up to 10 mg. The microgripper demonstrated handling capabilities for microcomponents such as pins in a breadboard (size: diameter 1 mm and length 10 mm; material: mild steel). Although this device exhibited acceptable handiness, it has the potential of handling numerous millimeter-scale components required in complex fabrication and assembly.

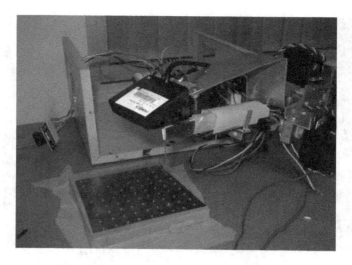

FIGURE 4.19 IPMC-based microgripper for handling objects [106].

4.4.3 AN IPMC-BASED ARTIFICIAL MUSCLE FINGER ACTUATED THROUGH EMG

Herein, an IPMC-based artificial muscle finger has been developed, which is actuated by the EMG signals generated from the movement of a human finger. The movement is sensed by an EMG sensor, which provides the signal for actuating the IPMC. When designing a microrobot/microgripper for grasping a lightweight object in an assembly, the IPMC-based finger behavior can be utilized to hold the object. During experimentation, the input parameter from muscle ranging from ±1.2 mV is taken through referenced single-ended (RSE) signal along with continuous pulse sample. These pulses are amplified with the help of a PXI system (amplification factor 2550). The desired output voltage range is generated through a DAC output port within the same frequency range. The output signal is connected to the IPMC strip. Due to the amplified output voltage from the DAC, an IPMC strip bends in one direction. By changing the polarity of the signal, the bending behavior can be reversed. The characteristics of the IPMC are traced on a graph paper for obtaining the deflection of the IPMC finger as shown in Figure 4.20. It shows that the IPMC gives a similar bending behavior as the human finger (Figure 4.12). The characteristics of the IPMC finger exhibits bending behavior at different voltages as given in Figures 4.21 and 4.22. By changing the movement of the finger in the opposite direction, the reverse behavior of the IPMC is obtained. It shows that the deflection changes with voltage by up to 12 mm. The major advantage of developing of an IPMC-based finger using EMG signals is that the tip of the IPMC finger follows the compliant behavior that can handle flexible components. The consumed energy for actuation is also less as compared to other conventional controllers. For this purpose, a stability analysis of EMG signals obtained from a human finger using a PID control system is

IPMCs as Soft Actuators

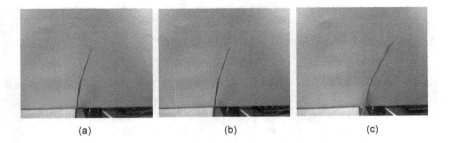

FIGURE 4.20 Movement of IPMC at different voltages [107].

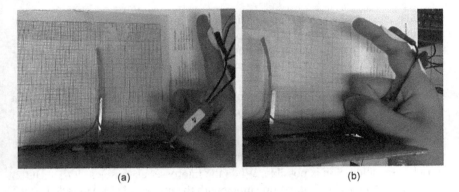

FIGURE 4.21 Control of IPMC by EMG signal in different states [107].

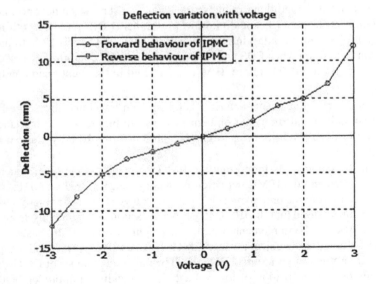

FIGURE 4.22 Deflection behavior of IPMC corresponding to voltage [107].

TABLE 4.5
Analysis of Different Conditions of Muscles [107]

	Intrinsic Muscles			Extrinsic Muscles					Polarity	
Cases	IO 1	IO 2	LU	EDC	EIP	FDS	FDP	State	Side A	Side B
1	OFF	OFF	OFF	OFF	OFF	OFF	OFF	None	None	None
2	ON	OFF	ON	OFF	OFF	ON	ON	Adduction	+ve	−ve
3	OFF	ON	ON	ON	ON	OFF	OFF	Abduction	−ve	+ve
4	ON	ON	ON	ON	ON	ON	ON	None	None	None

carried out because when an object is handled by an IPMC finger, the EMG signal provides the stability accordingly. By performing experiments, it is found that the IPMC finger achieves a similar movement of up to 12 mm. This potential is implemented as IPMC microfinger in the development of microrobot/microgripper for handling of micro-parts.

For the actuation of IPMC, the analysis of activated muscles is given in Table 4.5. We have analyzed different conditions during the contraction of different muscles of the index finger, such as intrinsic and extrinsic, which are responsible for the actuation of IPMC finger so that they can be used to hold an object. The two surfaces of the IPMC are denoted as side A and side B. In case of intrinsic muscles, the adduction is possible. When the interosseous muscles (IO 1 or IO 2) are in either "on" or "off" condition along with the lumbrical (LU) muscle in "on" condition, then it shows the abduction state. In case of extrinsic muscles, extensor digitorum communis (EDC) and extensor indicis proprius (EIP) muscles both have to be in "off" condition to achieve the adduction state when flexor digitorum superficialis (FDS) and flexor digitorum profundus (FDP) muscles both are in "on" condition, and for attaining the abduction state, EDC and EIP both have to be in "on" condition when FDS and FDP both are in "off" condition. In rest cases, no power is achieved. Therefore, IPMC is activated in the mentioned conditions of the muscles.

The voltage characteristic behavior is taken from EMG and fed to the IPMC for actuation in a real-time environment as shown in Figure 4.23. It is found that the trend of the IPMC actuation voltage is similar to the EMG voltage with an amplification factor.

From these data, it is found that the EMG frequency range is similar to the simulated data and the IPMC actuation frequency range is 48.5 ± 0.65 Hz, which is almost similar to the human muscle frequency range [107]. After these observations, it is understood that the IPMC finger behaves as an artificial muscle and this characteristic can be implemented in the development of an IPMC-based microgripper for holding the object through EMG. Thus, an IPMC strip can be applicable as a microfinger in micromanipulation. The major advantages of EMG-driven manipulators are utilization of low voltage, large bending amplitudes of IPMC and simple control toward the development of micro-/bio-robots.

IPMCs as Soft Actuators

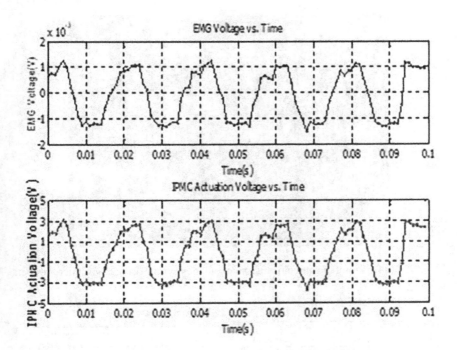

FIGURE 4.23 Different voltage responses in real-time environment [107].

4.4.4 Robotic Micro-Assembly Using IPMC Microgrippers

Herein, a multi-micromanipulation system has been developed, as shown in Figure 4.24. Multiple robotic assemblies were carried out with the help of four micromanipulators. Each micromanipulator was designed using a three-DOF manipulation system and an IPMC microgripper. The three-DOF manipulation system helps in manipulating the object in a three-dimensional workspace, and the IPMC microgripper helps in controlling the misalignment during peg-in-hole assembly. For attempting the robotic assembly in a large workspace, these IPMC-based microgrippers were integrated at each end effector position of micromanipulator. The micromanipulator was designed with two shaft mechanisms and one lead screw sliding mechanism to reach the identified location within workspace. The micro-gripping system was designed using IPMCs, where IPMCs are used as fingers for holding the object. For the demonstration of handling capability, a prototype of multi-micromanipulation system along with IPMC microgrippers is developed, which shows the pick-and-place and peg-in-hole assembly operations. Each micromanipulator exhibits two different kinds of trajectories during motion in different coordinates, as shown in Figure 4.25. This helps in performing the operations of a robotic micro-assembly.

The major advantages of this kind of multi-micromanipulation system using IPMC microgrippers are that a robotic micro-assembly can be performed the

FIGURE 4.24 Prototype of multi-micromanipulation system using IPMC microgrippers [108].

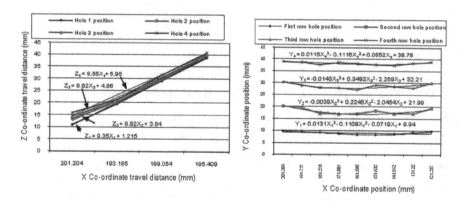

FIGURE 4.25 Performance of micromanipulator using IPMC microgripper for steel pegs [108].

operation of rigid and compliant objects. The IPMC soft actuators play vital roles because of their lightweight, large displacement, low mass force generation and misalignment compensation ability during micromanipulation. These capabilities are utilized for handling miniature parts such as pegs. The developed multi-micromanipulation system using IPMC microgrippers can handle pegs in a micro-assembly operation by shifting them from one hole to another hole in a large workspace (100 mm × 100 mm). By developing a prototype, it is demonstrated that IPMC-based microgrippers are capable of handling peg-in-hole assembly tasks in a multi-micromanipulation system.

4.5 CONCLUSIONS

In this book chapter, a development method for single-walled carbon nanotube (SWNT)-based IPMC soft actuators is discussed where the cost-effective hybrid-type sulfonated poly(1,4-phenylene ether-ether-sulfone) (SPEES) and functionalized single-walled carbon nanotubes (SWNT) based actuators produced by the film-casting method along with chemical reduction process of platinum ions as electrodes. Various chemical, material, mechanical and electromechanical properties were also discussed, which show the better performance as compared to other developed IPMC soft actuators. Further, the applications of IPMC soft actuators in various kinds of robotic assembly are described, which can be helpful for the development of a new generation of compliant mechanism using IPMC soft actuators. The major advantage of the IPMC-based soft actuator is that it can provide self-adjusting reaction forces by controlling the voltages of the actuators during assembly. This can also be useful in micro-assembly, which is a major demand in the industry. In biomimetic applications, the IPMC soft actuators can be helpful in replacing muscles and soft tissues and in cell manipulation, where they provide a solution toward non-damage of soft tissues of human body during manipulation. Hence, IPMC soft actuators have enormous potential for robotic and biomimetic applications in future.

ACKNOWLEDGMENT

The authors are grateful to the Director, CSIR-CMERI, Durgapur, India, for providing permission to publish this book chapter.

In the next chapter, the reader will find a new method of IPMC modeling (using Tractrix algorithm), where tip and shape estimation are shown with algorithm. This method is useful for robotic gripping applications.

REFERENCES

[1] M. Shahinpoor, Y. Bar-Cohen, J.O. Simpson and J. Smith, Ionic polymer metal composites (IPMC) as biomimetic sensors, actuators & artificial muscles- a review, *Smart Materials and Structures*, 7(6) (1998) R15, doi: 10.1088/0964-1726/7/6/001.
[2] M. Shahinpoor and K.J. Kim, The effect of surface-electrode resistance on the performance of ionic polymer-metal composite (IPMC) artificial muscle, *Smart Material and Structures*, 9(4) (2000) 543–551.
[3] M. Shahinpoor and K.J. Kim, Ionic polymer–metal composites: IV. Industrial and medical applications, *Smart Materials and Structures*, 14(1) (2004) 197–214, doi: 10.1088/0964-1726/14/1/020.
[4] R.K. Jain, S. Majumder and A. Dutta, SCARA based peg-in-hole assembly using compliant IPMC based micro gripper, *Robotics and Autonomous Systems*, 61(3) (2013) 297–311.
[5] R.K. Jain, S. Majumder and A. Dutta, Micro assembly by an IPMC based flexible 4-bar mechanism, *Smart Materials and Structures*, 21(7) (2012) 075004, doi: 10.1088/0964-1726/21/7/075004.

[6] R.K. Jain, S. Datta and S. Majumder, Bio-mimetic behavior of IPMC using EMG signal for a micro robot, *Mechanics Based Design of Structures and Machines: An International Journal*, 42(3) (2014) 398–417.

[7] X.L. Chang, P.S. Chee and E.H. Lim, A microreservoir-based drug delivery device using ionic polymer metal composite (IPMC) actuator, Proceedings of TENCON 2018-2018 IEEE Region 10 Conference, Jeju, Korea, 28–31 October, 2018, pp. 899–902.

[8] M.S. Saneei Mousavi, A.H. Karami, M. Ghasemnejad, M. Kolahdouz, F. Manteghi and F. Ataei, Design of a remote-control drug delivery implantable chip for cancer local on demand therapy using ionic polymer metal composite actuator, *Journal of the Mechanical Behavior of Biomedical Materials*, 86 (2018) 250–256, doi: 10.1016/j.jmbbm.2018.06.034.

[9] M. Shahinpoor and K. J Kim, Ionic polymer-metal composites: I. Fundamentals, *Smart Material Structure*, 10 (2001) 819–833.

[10] Nemat-Nasser and Y. Xian Wu, Comparative experimental study of ionic polymer-metal composites with different backbone ionomers and in various cation forms, *Journal of Applied Physics*, 93(9) (2003) 5255, doi: 10.1063/1.1563300.

[11] M. Billah, Z. M. Yusof, K. Kadir and A.M. MohdAli, Force measurement of non-linear IPMC using hydrodynamics analysis, IEEE 5th International Conference on Smart Instrumentation, Measurement and Application (ICSIMA), Songkla, Thailand, 28–30 November, 2018, doi: 10.1109/ICSIMA.2018.8688762.

[12] M. Shahinpoor, Continuum electromechanics of ionic polymeric gels as artificial muscles for robotic applications, *Smart Material Structure*, 3 (1994) 367, doi: 10.1088/0964–1726/3/3/012.

[13] M. Shahinpoor, Y. Bar-Cohen, T. Xue, J.O. Simpson and J. Smith, Ionic polymer-metal composites (IPMC) as biomimetic sensors and actuators-artificial muscles, Proceedings of SPIE's 5Ih Annual International Symposium on Smart Structures and Materials, San Diego, CA, 1–5 March, 1998, Paper No. 3324-27. pp. 1–21.

[14] M. Shahinpoor, Y. Bar-Cohen, T. Xue, J.O. Simpson and J. Smith, Ionic polymer-metal composites (IPMC) as biomimetic sensors and actuators-artificial muscles, ACS Book Series, Chapter XX, 1998, pp. 1–26.

[15] S. Nemat-Nasser and C. W. Thomas, Ionomeric polymer-metal composites, Chapter 6, In *Electroactive Polymer (EAP) Actuators as Artificial Muscles—Reality, Potential and Challenges*, Bar-Cohen (ed), SPIE, Bellingham, WA, 2001, pp. 139–191.

[16] G. Laurent and E. Piat, Efficiency of swimming micro robots using ionic polymer metal composites, IEEE International Conference on Robotics and Automation, Seoul Korea, 21–26 May, 2001, pp. 3914–3919.

[17] M. Shahinpoor and K. J. Kim, Experimental study of ionic polymer-metal composites in various cation forms: actuation behavior, *Science and Engineering of Composite Materials*, 10(6) (2002) 423–435, doi: 10.1515/SECM.2002.10.6.423.

[18] J. D. Madden, N. Vandesteeg, P. A. Anquetil, P. G. Madden, A. Takshi, R. Z. Pytel, S. R. Lafontaine, P. A. Wieringa and I. W. Hunter, Artificial muscle technology: physical principles and naval prospects, *IEEE Journal of Oceanic Engineering*, 29(3) (2004) 706–728.

[19] N.D. Bhat and W.J. Kim, Precision position control of ionic polymer metal composite, Proceeding of American Control Conference, Boston, Massachusetts, 30 June–2 July, 2004, pp. 740–745.

[20] Y. Bar-Cohen, Electroactive polymers as artificial muscles, Chapter 4, In *Compliant Structures in Nature and Engineering*, 20, 2005, pp. 69–81, doi: 10.2495/978-1-85312-941-4/04.

[21] S.J. Kim, H.I. Kim, S.J. Park, I.Y. Kim, S.H. Lee, T.S. Lee and S.I. Kim, Behavior in electric fields of smart hydrogels with potential application as bio-inspired actuators, *Smart Material Structure*, 14 (2005) 511, doi: 10.1088/0964-1726/14/4/008.
[22] B. Kim, D.H. Kim, J. Jung and J.O. Park, A biomimetic undulatory tadpole robot using ionic polymer–metal composite actuators, *Smart Material Structure*, 14 (2005) 1–7.
[23] S. G. Lee, H.C. Park, S.D. Pandita and Y. Yoo, Performance improvement of IPMC (ionic polymer metal composites) for a flapping actuator, *International Journal of Control, Automation and Systems*, 4(6) (2006) 748–755.
[24] S. Lee and K.J. Kim, Muscle-like linear actuator using an ionic polymer-metal composite and its actuation characteristics, *Smart Structures and Materials 2006: Electroactive Polymer Actuators and Devices (EAPAD)*, 6168 (2006) 616820, doi: 10.1117/12.655257.
[25] C.K. Chung, P.K. Fung, Y.Z. Hong, M.S. Ju, C.C.K. Lin and T.C. Wu, A novel fabrication of ionic polymer-metal composites (IPMC) actuator with silver nanopowders, *Sensors and Actuators B*, 117 (2006) 367–375.
[26] E. Malone and H. Lipson, Freeform fabrication of ionomeric polymer-metal composite actuators, *Rapid Prototyping Journal*, 12(5) (2006) 484–502.
[27] K.J. Kim, W. Yim, J.W. Paquette and D. Kim, Ionic polymer metal composites for underwater operation, *Journal of Intelligent Material Systems and Structures*, 18(2) (2007) 123–131.
[28] B.K. Fang, M.S. Ju and C.C.K. Lin, A new approach to develop ionic polymer–metal composites (IPMC) actuator: fabrication and control for active catheter systems, *Sensors and Actuators A*, 137 (2007) 321–329.
[29] M. Konyo, S. Tadokoro and K. Asaka, Applications of ionic polymer-metal composites: multiple-DOF devices using soft actuators and sensors, In *Electroactive Polymers for Robotic Applications*, Kim K.J., Tadokoro S. (eds), Springer, London, 2007, doi: 10.1007/978-1-84628-372-7_9.
[30] D. Pugal, *Model of a Self-Oscillating Ionic Polymer-Metal Composite Bending Actuator*, Master Thesis, Tartu University, Estonia, 2008.
[31] A. Ji, H.C. Park, Q.V. Nguyen, J.W. Lee and Y.T. Yoo, Verification of beam models for ionic polymer-metal composite actuator, *Journal of Bionic Engineering*, 6 (2009) 232–238, doi: 10.1016/S1672-6529(08)60117-1.
[32] Z. Chen and X. Tan, A control-oriented and physics-based model for ionic polymer–metal composite actuators, *IEEE/ASME Transactions on Mechatronics*, 13(5) (2008) 519–529.
[33] Z. Chen, D.R. Hedgepeth and X. Tan, A nonlinear, control-oriented model for ionic polymer–metal composite actuators, *Smart Material and Structure*, 18 (2009) 055008 (9pp), doi: 10.1088/0964-1726/18/5/055008.
[34] Z. Chen, *Ionic Polymer Metal Composite Artificial Muscles and Sensors: A Control Systems Perspective*, Ph.D. Thesis, Michigan State University, USA, 2009.
[35] Z. Chen, T.I. Um and H.B. Smith, Bio-inspired robotic manta ray powered by ionic polymer–metal composite artificial muscles, *International Journal of Smart and Nano Materials*, 3(4) (2012) 296–308, doi: 10.1080/19475411.2012.686458.
[36] Z. Chen, T. Um and H.B. Smith, Ionic polymer-metal composite artificial muscles in bio-inspired engineering research: underwater propulsion, Chapter 10, In *Smart Actuation and Sensing Systems–Recent Advances and Future Challenges*, 2012, pp. 223–248. doi: 10.5772/51292.
[37] J.H. Jung, V. Sridhar and I.K. Oh, Electro-active nano-composite actuator based on fullerene-reinforced Nafion, *Composites Science and Technology*, 70(4) (2010) 584–592.

[38] S. Saher, S. Moon, S.J. Kim, H.J. Kim and Y.H. Kim, O2 plasma treatment for ionic polymer metal nano composite (IPMNC) actuator, *Sensors and Actuators B: Chemical*, 147 (2010) 170–179.

[39] Y. Bahramzadeh and M. Shahinpoor, Characterizing of ionic polymer-metal composites (IPMC) for sensitive curvature measurement, ASME 2010 Conference on Smart Materials, Adaptive Structures and Intelligent Systems, Philadelphia, Pennsylvania, USA, September 28–October 1, 2010, SMASIS2010-3799, pp. 269–274, 6 pages, doi: 10.1115/SMASIS2010-3799.

[40] S. Mukherjee and R. Ganguli, Ionic polymer metal composite flapping actuator mimicking Dragonflies, *CMC*, 19(2) (2010) 105–133.

[41] S. Gutta, *Modeling and Control of a Flexible Ionic Polymer Metal Modeling and Control of a Flexible Ionic Polymer Metal Composite (IPMC) Actuator for Underwater Propulsion Composite (IPMC) Actuator*, Ph.D. Thesis, University of Nevada, Las Vegas, USA, 2011.

[42] V. Panwar, C. Lee, S.Y. Ko, J.O. Park and S. Park, Dynamic mechanical, electrical, and actuation properties of ionic polymer metal composites using PVDF/PVP/PSSA blend membranes, *Materials Chemistry and Physics*, 135 (2012) 928–937.

[43] A.D. Price, *Fabrication, Modelling and Application of Conductive Polymer Composites*, Ph. D. Thesis, University of Toronto, Canada, 2012.

[44] D. Lantada, P.L. Morgado, J.L. Muñoz Sanz, J.M. Muñoz Guijosa and J. Echávarri Otero, Neural network approach to modeling the behavior of ionic polymer-metal composites in dry environments, *Journal of Signal and Information Processing*, 3 (2012) 137–145, doi: 10.4236/jsip.2012.32018.

[45] J.S. Lee, W. Yim, C. Bae and K.J. Kim, Wireless actuation and control of ionic polymer metal composite actuator using a microwave link, *International Journal of Smart and Nano Materials*, 3(4) (2012) 244–262, doi: 10.1080/19475411.2012.670141.

[46] Q. He, M. Yu, Y. Li, Y. Ding, D. Guo and Z. Dai, Investigation of ionic polymer metal composite actuators loaded with various tetraethyl orthosilicate contents, *Journal of Bionic Engineering*, 9(1) (2012) 75–83.

[47] S. Kang, W. Kim, H.J. Kim and J. Park, Adaptive feed forward control of ionic polymer metal composites with disturbance cancellation, *Journal of Mechanical Science and Technology*, 26(1) (2012) 205–212, doi: 10.1007/s12206-011-0916-8.

[48] L. Shi, S. Guo and K. Asaka, Modeling and experiments of IPMC actuators for the position precision of underwater legged microrobots, Proceeding of the IEEE International Conference on Automation and Logistics, Zhengzhou, China, August, 2012, pp. 420–425.

[49] K.K. Leang, Precision control of ionic polymer-metal composite actuators, The 7th World Congress on Biomimetics, Artificial Muscles and Nano-Bio (BAMN2013), Jeju Island, South Korea, 26–30 August, 2013.

[50] R. Caponetto, S. Graziani, F. Sapuppo and V. Tomasello, An enhanced fractional order model of ionic polymer-metal composites actuator, *Advances in Mathematical Physics*, 2013 Article ID 717659 (2013), 6 pages, doi: 10.1155/2013/717659.

[51] V. D. Luca, P. Digiamberardino, G. D. Pasquale, S. Graziani, A. Pollicino, E. Umana and M.G. Xibilia, Ionic electroactive polymer metal composites: fabricating, modeling, and applications of post silicon smart devices, *Journal of Polymer Science, Part B: Polymer Physics*, 51 (2013) 699–734.

[52] M. Sasaki, W. Lin, H. Tamagawa, S. Ito and K. Kikuchi, Self-sensing control of Nafion-based ionic polymer-metal composite (IPMC) actuator in the extremely low humidity environment, *Actuators*, 2 (2013) 74–85, doi:10.3390/act2040074.

[53] H. Okazaki, S. Sawada, T. Matsuda and M. Kimura, Soft actuator using ionic polymer-metal composite driven with ionic liquid, IEEE International Meeting for Future of Electron Devices, Kansai (IMFEDK), Kyoto, Japan, 19–20 June, 2014, pp. 1–2, doi: 10.1109/IMFEDK.2014.6867092.
[54] X. Chen, Adaptive control for ionic polymer-metal composite actuator based on continuous-time approach, Proceedings of the 19th World Congress The International Federation of Automatic Control, Cape Town, South Africa, August 24–29, 2014, pp. 5073–5078.
[55] J. Kwaśniewski, I. Dominik and F. Kaszuba, Energy harvesting system based on ionic polymer-metal composites-identification of electrical parameters, *Polish Journal of Environmental Studies*, 23(6) (2014) 2339–2343.
[56] M. Farid, Z. Gang, T. L. Khuong and Z. Z. Sun, Forward kinematic modeling and simulation of ionic polymer metal composites (IPMC) actuators for bionic knee joint, *Advanced Materials Research*, 889–890 (2014) 938–941.
[57] I. Dominik, F. Kaszuba and J. Kwaśniewski, Modelling coupled electric field and motion of beam of ionic polymer-metal composite, *Acta Mechanica et Automatica*, 8(1) (2014) 38–43, doi: 10.2478/ama-2014-0007.
[58] R. Mutlu, G. Alici, X. Xiang and W. Li, An active-compliant micro-stage based on EAP artificial muscles, IEEE/ASME International Conference on Advanced Intelligent Mechatronics (AIM), Besançon, France, 8–11 July, 2014, pp. 611–616.
[59] Y. Tang, C. Chen, Y.S. Ye, Z. Xue, X. Zhou and X. Xie, The enhanced actuation response of an ionic polymer–metal composite actuator based on sulfonated polyphenylsulfone, *Polymer Chemistry*, 5 (2014) 6097–6107.
[60] V. Palmre, D. Pugal, K.J. Kim, K.K. Leang, K. Asaka and A. Aabloo, Nanothorn electrodes for ionic polymer-metal composite artificial muscles, *Scientific Reports*, 4 (2014) 6176 (1–10), doi: 10.1038/srep06176.
[61] L.A. Hirano, L.W. Acerbi, K. Kikuchi, S. Tsuchitani and C.H. Scuracchio, Study of the influence of the hydration level on the electromechanical behavior of Nafion based ionomeric polymer-metal composites actuators, *Materials Research*, 18 (2015) 4, doi: 10.1590/1516-1439.353214.
[62] A. Simaite, Development of ionic electroactive actuators with improved interfacial adhesion: towards the fabrication of inkjet printable artificial muscles, *Micro and Nanotechnologies/Microelectronics*, INSA de Toulouse, 2015. English. NNT: 2015ISAT0043.
[63] B. Sun, D. Bajon, J. M. Moschetta, E. Benard and C. Thipyopas, Integrated static and dynamic modeling of an ionic polymer–metal composite actuator, *Journal of Intelligent Material Systems and Structures*, 26(10) (2015) 1164–1178, doi: 10.1177/1045389X14538528.
[64] J.D. Carrico, N.W. Traeden, M. Aureli and K.K. Leang, Fused filament 3D printing of ionic polymer-metal composites (IPMCs), *Smart Material Structure*, 24 (2015) 125021 (11 pp), doi: 10.1088/0964-1726/24/12/125021.
[65] J. Nam, *Ionic Polymer-Metal Composite Actuators Based on Nafion Blends with Functional Polymers*, Master of Science Thesis, University of Nevada, Las Vegas, USA, 2016.
[66] Q. Shen, S. Trabia, T. Stalbaum, V. Palmre, K. Kim and I.K. Oh, A multiple-shape memory polymer metal composite actuator capable of programmable control, creating complex 3D motion of bending, twisting, and oscillation, *Scientific Reports*, 6 (2016) 24462 (pp. 1–11), doi: 10.1038/srep24462.
[67] W. Hong, A. Almomani and R. Montazami, Electrochemical and morphological studies of ionic polymer metal composites as stress sensors, *Measurement*, 95 (2017) 128–134, doi: 10.1016/j.measurement.2016.09.036.

[68] S.D. Peterson and M. Porfiri, Energy exchange between coherent fluid structures and ionic polymer metal composites, toward flow sensing and energy harvesting, Chapter 14, In *Ionic Polymer Metal Composites (IPMCs): Smart Multi-Functional Materials and Artificial Muscles*, M. Shahinpoor (ed), 2, 2016, pp. 1–18, doi: 10.1039/9781782627234.
[69] M. Haq and Z. Gang, Ionic polymer–metal composite applications, *Emerging Materials Research*, 5(1) (2016) 153–164, doi: 10.1680/jemmr.15.00026.
[70] A. Kazem and J. Khawwaf, Estimation bending deflection in an ionic polymer metal composite (IPMC) material using an artificial neural network model, *Jordan Journal of Mechanical and Industrial Engineering*, 10(2) (2016) 123–131.
[71] K. Bian, H. Liu, G. Tai, K. Zhu and K. Xiong, Enhanced actuation response of Nafion-based ionic polymer metal composites by doping BaTiO3 nanoparticles, *Journal of Physics and Chemistry*, 120(23) (2016) 12377–12384.
[72] Z. Chen, A review on robotic fish enabled by ionic polymer–metal composite artificial muscle, *Robotics and Biomimetics*, 4(24) (2017) 1–13, doi: 10.1186/s40638-017-0081-3.
[73] G.D. Pasquale, S. Graziani, C. Gugliuzzo and A. Pollicino, Ionic polymer-metal composites (IPMCs) and ionic polymer-polymer composites (IP2Cs): effects of electrode on mechanical, thermal and electromechanical behavior, *AIMS Materials Science*, 4(5) (2017) 1062–1077, doi: 10.3934/matersci.2017.5.1062.
[74] J.D. Carrico, T. Tyler and K.K. Leang, A comprehensive review of select smart polymeric and gel actuators for soft mechatronics and robotics applications: fundamentals, freeform fabrication, and motion control, *International Journal of Smart and Nano Materials*, 8(4) (2017) 144–213, doi: 10.1080/19475411.2018.1438534.
[75] M. Porfiri, H. Sharghi and P. Zhang, Modeling back-relaxation in ionic polymer metal composites: the role of steric effects and composite layers, *Journal of Applied Physics*, 123 (2018) 014901, doi: 10.1063/1.5004573.
[76] A. Boldini, K. Jose, Y. Cha and M. Porfiri, Enhancing the deformation range of ionic polymer metal composites through electrostatic actuation, *Applied Physics Letter*, 112 (2018) 261903, doi: 10.1063/1.5037889.
[77] J. Bernat and J. Kolota, Integral multiple models online identifier applied to ionic polymer–metal composite actuator, *Journal of Intelligent Material Systems and Structures*, 29 (14) (2018) 2863–2873, doi: 10.1177/1045389X18781027.
[78] T.L. Sainag and S. Mukherjee, Optimal position control of ionic polymer metal composite using particle swarm optimization, *Electroactive Polymer Actuators and Devices (EAPAD) XX*, 105941D (2018), doi: 10.1117/12.2284627.
[79] Y. Wang, J. Liu, D. Zhu and H. Chen, Active tube-shaped actuator with embedded square rod-shaped ionic polymer-metal composites for robotic-assisted manipulation, *Applied Bionics and Biomechanics*, 2018 (Article ID 4031705) (2018) 12 pages, doi: 10.1155/2018/4031705.
[80] Y. Zhao, D. Xu, J. Sheng, Q. Meng, D. Wu, L. Wang, J. Xiao, W. Lv, Q. Chen and D. Sun, Biomimetic beetle-inspired flapping air vehicle actuated by ionic polymer-metal composite actuator, *Applied Bionics and Biomechanics*, 2018 (Article ID 3091579) (2018) 7 pages, doi: 10.1155/2018/3091579.
[81] M. Shahinpoor and E. Tabatabaie, Soft biomimetic robotic looped haptic feedback sensors, *Journal of Robotics Engineering and Automation Technology (JREAT-102)*, 2018(01) (2018) 1–10, doi: 10.29011/JREAT-102/100002.
[82] A. Khan, Inamuddin, R.K. Jain and M. Naushad, Fabrication of a silver nano powder embedded Kraton polymer actuator and its characterization, *RSC Advances*, 5 (2015) 91564–91573, doi: 10.1039/c5ra17776f, Impact Factor: 3.049.
[83] A. Khan, R.K. Jain, B. Ghosh, Inamuddin and A.M. Asiri, Novel ionic polymer–metal composite actuator based on sulfonated poly (1,4-phenylene ether ether-sulfone)

and polyvinylidene fluoride/ sulfonated graphene oxide, *RSC Advances*, 8 (2018) 25423–25435.
[84] D.K. Biswal and P. Padhi, Experimental investigation on the bending response of multilayered Ag-ionic polymer metal composite actuator for robotic application, *Advanced Materials Letters*, 9(8) (2018) 544–548, doi: 10.5185/amlett.2018.1753.
[85] D.K. Biswal, D. Bandopadhya and S.K. Dwivedy, A non-linear dynamic model of ionic polymer-metal composite (IPMC) cantilever actuator, *International Journal of Automotive and Mechanical Engineering*, 16(1) (2019) 6332–6347.
[86] E.S. Tabatabaie and M. Shahinpoor, Novel configurations of slit tubular soft robotic actuators and sensors made with ionic polymer metal composites (IPMCs), *Robotics and Automation Engineering Journal*, 3(4) (2018) 90–99, doi: RAEJ.MS.ID.5555616.
[87] E.S. Tabatabaie, *Novel Configurations of Ionic Polymer-Metal Composites (IPMCs) As Sensors, Actuators, and Energy Harvesters*, Ph. D, University of Maine, USA, 2019.
[88] E.S. Tabatabaie and M. Shahinpoor, Artificial soft robotic elephant trunk made with ionic polymer-metal nanocomposites (IPMCs), *International Journal of Robotics and Automation*, 5(4) (2019) 138–142.
[89] J. Zhang, J.S. Ciaran T.O. Neill, C.J. Walsh, R.J. Wood, J.H. Ryu, J.P. Desai and M.C. Yip, Robotic artificial muscles: current progress and future perspectives, *IEEE Transactions on Robotics*, 35(3) (2019) 7611–7781, doi: 10.1109/TRO.2019.2894371.
[90] A. Gupta and S. Mukherjee, Dynamic modeling of biomimetic undulatory underwater propulsor actuated by ionic polymer metal composites, World Congress on Advances in Structural Engineering and Mechanics (ASEM19), Jeju Island, Korea, September 17–21, 2019.
[91] J. Wang, Y. Wang, Z. Zhu, J. Wang, Q. He and M. Luo, The effects of dimensions on the deformation sensing performance of ionic polymer-metal composites, *Sensors*, 19 (2019) 2104 (1–12), doi: 10.3390/s19092104.
[92] W.H. MohdIsa, A. Hunt and S. Hassan HosseinNia, Sensing and self-sensing actuation methods for ionic polymer–metal composite (IPMC): a review, *Sensors*, 19 (2019) 3967 (1–37), doi: 10.3390/s19183967.
[93] M. Annabestani and M. Fardmanesh, Ionic Electro active Polymer-Based Soft Actuators and their Applications in Microfluidic Micropumps, Microvalves, and Micromixers: A Review, arXivpreprint arXiv:1904.07149, 2019.
[94] A.S. Tripathi, B.P. Chattopadhyay and S. Das, Cost-effective fabrication of ionic polymer based artificial muscles for catheter-guidewire maneuvering application, *Microsystem Technologies*, 25 (2019) 1129–1136, doi: 10.1007/s00542-018-4152-3.
[95] S. Li and J. Yip, Characterization and actuation of ionic polymer metal composites with various thicknesses and lengths, *Polymers (Basel)*, 11(01) (2019) 91 (1–14), doi: 10.3390/polym11010091.
[96] R. Gonçalves, K.A. Tozzi, M.C. Saccardo, A.G. Zuquello and C.H. Scuracchio, Nafion-based ionomeric polymer/metal composites operating in the air: theoretical and electrochemical analysis, *Journal of Solid State Electrochemistry*, (2020) 1845–1856, doi: 10.1007/s10008-020-04520-6.
[97] L. Yang, D. Zhang, X. Zhang and A. Tian, Prediction of the actuation property of Cu ionic polymer–metal composites based on backpropagation neural networks, *ACS Omega*, 5 (2020) 4067–4074.
[98] S. Ma, Y. Zhang, Y. Liang, L. Ren, W. Tian and L. Ren, High-performance ionic-polymer–metal composite: toward large-deformation fast-response artificial muscle, *Advanced Functional Materials*, 30 (2020) 1908508 (1–9), doi: 10.1002/adfm.201908508.
[99] A. Khan, R.K. Jain, P. Banerjee, B. Ghosh, Inamuddin and A.M. Asiri, Development, characterization and electromechanical actuation behavior of ionic

polymer metal composite actuator based on sulfonated poly(1,4-phenylene ether-ether-sulfone)/carbon nanotubes, *Scientific Reports*, 8 (2018) 9909 (1–16), doi: 10.1038/s41598-018-28399-6.

[100] Inamuddin, A. Khan, R.K. Jain and M. Naushad, Study and preparation of highly water-stable polyacrylonitrile-Kraton-graphene composite membrane for bending actuator toward robotic application, *Journal of Intelligent Materials System and Structure*, 27 (2016) 1534–1546.

[101] Inamuddin, A. Khan, R.K. Jain and M. Naushad, Development of sulfonated poly(vinyl alcohol)/polypyrrole based ionic polymer metal composite (IPMC) actuator and its characterization, *Smart Materials and Structures*, 24 (2015) 95003.

[102] M. Rajagopalan, J.H. Jeon and I.K. Oh, Electric-stimuli-responsive bending actuator based on sulfonated polyetherimide, *Sensors Actuators B Chemical*, 151 (2010) 198–204.

[103] H. Lian et al., Enhanced actuation in functionalized carbon nanotube-Nafion composites, *Sensors Actuators B Chemical*, 156 (2011) 187–193.

[104] T.A. Zawodzinski, Water uptake by and transport through Nafion® 117 membranes. *Journal of Electrochemical Society*, 140 (1993) 1041.

[105] R.K. Jain, U.S. Patkar and S. Majumder, Micro gripper for micro manipulation using IPMCs, *Journal of Scientific & Industrial Research*, 68 (2009) 23–28.

[106] R.K. Jain, S. Datta, S. Majumder and A. Dutta, Two IPMC fingers based micro gripper for handling, *International Journal of Advanced Robotics Systems*, 8(1) (2011) 1–9, doi: 10.5772/10523.

[107] R.K. Jain, S. Datta and S. Majumder, Design and control of an IPMC artificial muscle finger for micro gripper using EMG signal, *Mechatronics*, 23(3) (2013) 381–394, doi: 10.1016/j.mechatronics.2013.02.008.

[108] R.K. Jain, S. Datta, S. Majumder and A. Dutta, Development of multi micro manipulation system using IPMC micro grippers, *Journal of Intelligent and Robotics Systems*, 74(3) (2014) 547–569, doi: 10.1007/s10846-013-9874-y.

5 Inverse Kinematic Modeling of Bending Response of Ionic Polymer Metal Composite Actuators

Siladitya Khan
University of Rochester

Gautam Gare
Carnegie Mellon University

Ritwik Chattaraj
Indian Institute of Engineering Science and Technology

Srijan Bhattacharya
RCC Institute of Information Technology

Bikash Bepari
Haldia Institute of Technology

Subhasis Bhaumik
Indian Institute of Engineering Science and Technology

CONTENTS

5.1 Introduction 96
5.2 Soft Robotic Materials: Research Status and Current Trends 97
5.3 Simulating Soft Actuators: A Robotic Perspective 98
    5.3.1 Early Developmental Efforts and Modelling Challenges 99
    5.3.2 Hyper-Redundant Serial Chain Approximation of Soft
          Robotic Actuators 100
    5.3.3 The Jacobian Transpose and Pseudo-Inverse Solution to
          Inverse Kinematics 102

DOI: 10.1201/9781003204664-5

5.3.4   'Tractrix'-Based Solution to Inverse Kinematics 103
         5.3.4.1   Experimental Validation by Simulating Doped
                   IPMC Actuators 106
5.4   Design and Development of Compliant Soft Actuated Grippers: An
      Application of Hyper-Redundant Kinematic Modelling 113
5.5   Conclusions and Future Prospects 114
Acknowledgements 117
References 117

5.1   INTRODUCTION

The inspiration for the development of autonomous soft robotic mechanisms bearing close affinity to biological systems is often derived from nature's foray. The research affinity is drawn towards the identification of compatible materials for designing diverse mechanisms and structures, while addressing umpteen challenges in robustness and versatility. In contrast to conventional robotic systems that are characterized by highly rigid components, soft robots are saliently differentiable by their elastic and flexible nature while still retaining their transductional capabilities. The mechanical transformation and reconfiguration is achieved by involving materials such as polymers, plastics, fluids and colloids that undergo phase transition to exhibit significant deformative behaviours [1–3]. Such materials are driven by diverse action stimuli such as humidity, electric field, heat and magnetic intervention, change in pH or solvent concentration. In order to address diverse research challenges encountered in the domain of mechatronics and robot development, a comprehensive analysis of soft actuators–sensors in terms of their material deployments, modelling strategies and engaged applications is anticipated. While their use is set to change the way the conventional design problems are approached, a significant motivation in utilizing the basic robotic concepts for operational modelling is liable to be drawn.

Despite the multifaceted advantages offered by soft robots, their development faces significant challenges in design selection, modelling and control. Unlike the conventional robotic set-ups whose configurations are usually controlled by well-structured mathematical foundations, soft robotic assemblies are driven by complex partial differential sets. Thus, an efficient modelling scheme capable of simulating actuator behaviour, at both the material and geometric levels, is anticipated. In the quest to address the issue, several modelling alternatives have been proposed, ranging from the ones adroit to explain minute molecular mechanisms to parameter-driven system simulations. Soft robots are known for their capabilities to continually adapt their trajectories to incessant changes in body shape and structure. This phenomenon has motivated a new paradigm of mechanics-based smart material approximations, through which continuum structures bereft of discrete joint linkages are approximated as serial hyper-redundant chains driven by kinematic frameworks. The piece-wise curvature estimation method presents an alternate narrative to multi-physics-based electromechanical or chemical deliberations [4–6]. The properties of smart materials once captured in the model frameworks are anticipated for their potential deployments in diverse paradigms.

The continual rise of robot use in unstructured environments and precision applications has facilitated the deployment of deformable and variable stiffness technologies to build new classes of robotic systems. The applications are distributed across broad areas of mechanical, aerospace, biomedical and electrical engineering for establishing safer and natural human interactions while accomplishing uncertain and dynamic tasks [7–9]. The concepts of actuation as well as sensing in soft robotic avatars are still an emerging field of investigation that are under the research scrutiny. Therefore, deliberations emphasizing on the pertinent research problems such as the ones discussed in the present chapter are anticipated to guide both practitioners and freshmen to acquire a better understanding of fundamental concepts, observational requirements and relevant tools for pursuing future investigations.

This chapter offers an overview of modelling techniques that are used to explain transductional phenomenon of smart materials. The study is expected to guide through various alternatives in terms of available soft materials that either are used, or bear potential in robotic applications. The specific domain of kinematics-based analysis of soft polymer actuators is presented in this investigation, which shows the immense prospects of their deployment in various applications. A specific case study of compliant ionic polymer-based gripper is undertaken to validate the utility of the modelling methodology.

5.2 SOFT ROBOTIC MATERIALS: RESEARCH STATUS AND CURRENT TRENDS

Soft robotic materials can be either synthetic or natural in their material composition and act by transforming the embodied chemical or physical energy into mechanical work. Their action is primarily guided by changes in various parameters such as heat, electric or magnetic potential, humidity, concentration differential and change in ambient lighting. The prototypical 'soft muscle' has evolved from its original bio-counterpart comprising of adenosine triphosphate (ATPase) enzyme as the energy carrier for the cardiomyocytes. Recent efforts have been aimed at bioengineering, such as organic substrates to design miniature bio-actuators and micro-devices. Artificially cultured cardiomyocytes are utilized to convert the biochemical energy into mechanical energy in a microchip to design pulsatile micro-actuators. Elastomers were developed and identified early, to demonstrate actuation behaviour, and the same had been classified into primarily two classes, the first being electro-responsive type and the other being photo-responsive type. The electro-responsive type of elastomers such as polyurethane demonstrated the phenomenon of electrostriction, under the action of which the sample exhibited high deformations of its compliant body. A common photo-responsive elastomer is a liquid crystalline film, which can be bent precisely in any direction by the action of a linearly polarized light beam. Baughman and his associates [23] reported a novel electromechanical actuator known as carbon nanotubes (CNTs), which demonstrated quantum chemical expansions as a result of its charging electrochemical double layer. Polymer hydrogels have been reported to exhibit actuation characteristics over a wide range of dominant stimuli such as pH, salt, solvent and electrical induction. Conductive polymers

(CPs), such as polypyrrole, polythiophene and their derivatives, show dimensional changes when subjected to electrochemical doping. Interesting investigations have resulted in humidity-responsive CPs, which revealed the presence of electrochemically synthesized polypyrrole films that undergo rapid bending due to water vapour absorption.

Most of the materials that have been discussed until now involve 'dynamic' changes of morphology, primarily tip or whole body deflection and torsional twists about the neutral. However, a certain class of materials exhibit a rather radical approach in transformation manifested as 'plastic' changes, which involves a complete change in body shape. In the parlance of robotics, plastic changes in bodily morphologies have long been investigated in the context of reconfigurable robotics. The robots of such type are known to be built of smaller modules being capable of remaining connected or disconnected from each other. Modularity and reconfigurability of such type have been replicated by foams and colloids that undergo chemical synthesis within their constituent molecules to represent changes in structural design. Despite the large variety in actuation patterns, deflection or bending corresponds to the most common form of behaviour noticeable among smart materials. Table 5.1 provides a brief overview of the various soft actuation materials under the influence of their corresponding stimuli along with their year of investigation. The primary aim of the list is to provide a concise summary to the young, yet stimulating research field with a brief introduction of enabling technologies and conceptual issues. In a broader light, this chapter aims to identify the important developments and challenges towards the development of soft robots.

Among the various soft actuator materials reported above, electrically driven polymer actuators, commonly referred to as electroactive polymer (EAP) actuator, are of foremost interest owing to their typical characteristics such as low driving voltage, high deflection and their ability to be activated in hydrated as well as dry states. EAPs have gained a considerable mileage in use over their counterparts and have found diverse application interests [49–52]. Conventional actuation units such as motors are being increasingly phased out by modern soft alternatives that are mechanically biocompatible and capable of demonstrating lifelike functionalities.

5.3 SIMULATING SOFT ACTUATORS: A ROBOTIC PERSPECTIVE

Besides exploring newer alternatives in terms of availability of physical materials for design and development of soft robots, simulation and modelling play a significant role for a systematic and holistic development of this research field. In particular, newer alternatives are required that are able to rise above the conventional chemical orientations. Owing to the diversity in principles governing actuation schemes across various materials, there is a lack of a uniform approach that is capable of guiding geometric manifestations of actuation behaviour. Concurrently, the deployment of such actuation elements in the domain of robotics

Time-Dependent Bending of IPMC Actuators

TABLE 5.1
Characteristics of Various Soft Robotic Elements

Serial	Material	Stimuli	Action	Year	Source
1	Biological muscle bundles	ATP concentration	Elongation/contraction	2000	[10–13]
2	Elastomers	Light intensity variation/electric potential	Bending/extension–relaxation	1998	[14–22]
3	Carbon nanotubes	Electric potential	Bending/torsion	2005	[23–26]
4	Foams/colloids	Chemical synthesis	Shape reconfiguration	2014	[27]
5	Polymers (thermosetting/thermoplastic/electroactive)	Humidity/electric potential	Bending deflections/torsions/contraction–extension	1998	[28–37]
6	Polymer hydrogels	Magnetic or electric potential/light intensity/solvent concentration/pH	Flow	2000	[38–40]
7	Shape-memory alloys	Heat	Bending/torsion	1990	[41–44]
8	Piezoelectric composites	Electric potential	Bending deflections	1985	[45]
9	Pneumatic muscles	Pneumatic flow	Bending/torsion	1995	[46]
10	Polymeric acids	Concentration change	Swelling/de-swelling	1949	[47]
11	ER/MR fluids	Shear/normal stress	Flow	1989	[48,49]

has called for a mechanics-based method that is able to represent the physical phenomenon without engaging in derivations from the chemical perspective. This chapter provides an understanding of the available tools in the domain of robotics that are mechanics based and may be employed to simulate material behaviour. The techniques discussed in the present section are directed at representing the soft flexible body of such actuators in terms of a collection of a few discretized links. The deflection patterns are subsequently sought to be emulated by resorting to inverse kinematics (IK)-based procedures. The various IK routines that have been deployed in this context have subsequently been discussed, and finally, a specific application of such a modelling approach is presented in the form of a design of a compliant jaw microgripper. The role of simulation in performing its workspace analysis and load-carrying capability is highlighted.

5.3.1 Early Developmental Efforts and Modelling Challenges

Prior efforts directed at developing an efficient model capable of dictating the deformation of a smart actuator have been established, in the form of methodologies that are adept at estimating solely the end effector movements. Such a modelling approach suffices in scenarios where mere convergence between the model and the actuator at the distal end solves the problem. However, under conditions where the overall profile plays a crucial role for specific applications, a foresight of the actuator's dynamic curvature needs to be ascertained. The same is accomplished by developing a reliable methodology towards emulating the complete actuator behaviour. Significant efforts have been invested in developing a modelling philosophy for such type of actuators employing mechanical beam theories. In this context, two notable propositions are worth mentionable, the first being the Euler–Bernoulli model and the other being Timoshenko's beam theory. The Euler–Bernoulli model is founded upon the assumption of a strict perpendicular orientation between the central axis of the beam and its end plate, which holds good for elements whose diameter is negligible compared to its length. However, for shorter beams, it fails to be adequately representative; under such instances, Timoshenko's approach is adopted that yields better approximations for shorter beam elements. The approach had been adopted to estimate the deflection behaviour of a bi-layer polymer actuator and a tri-layer EAP. The radius of the curvature generated as a result of deflection was measured, from which the developed strain was calculated [53–56]. Such efforts were refined taking the non-linear actuation patterns into account [57]; however, all such methodologies were restrained by the assumption of a constant modulus of elasticity, when in actuated state. Additionally, the deflection equations are increasingly rendered complex for composite layered architectures over a simplistic single solid beam model. In order to mitigate such complications, a novel approach of mimicking flexible actuator behaviour was accomplished by approximating it as a collection of serially linked elements. The redundancy of the complete structure was resolved using an IK-based procedure known as the angle OPT method [58]. However, since the event was viewed as a constrained optimization problem, it required case-specific angular assignments in order to ensure that the model converges to the tracked polymer feature. The primary difficulty that arises from such a formulation lies in specific joint space constraints over a generic kinematic architecture to characterize actuator motion under different experimental conditions.

5.3.2 Hyper-Redundant Serial Chain Approximation of Soft Robotic Actuators

The present discussion explores the paradigm of soft actuator modelling using a kinematics-based procedure for effective simulation of the actuator profile. It is, however, anticipated to refrain from assigning additional restrictions on the joint movements through boundary value allocations, in order to achieve convergence between the tracked and simulated actuator postures. The complete actuator element is approximated as a serial chain manipulator comprised of discretized links connected by revolute joints as shown in Figure 5.1. Generally, a greater number of links can ensure a more accurate imitation of its flexible nature, which in turn

Time-Dependent Bending of IPMC Actuators

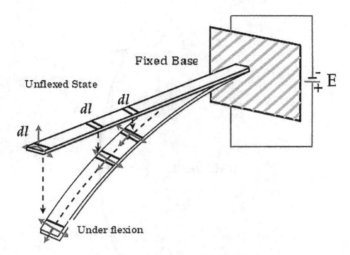

FIGURE 5.1 Approximation of soft actuator as discrete multi-linked planar manipulator.

becomes responsible for increasing the computational cost while resolving the IK solution of the same. The mathematical relationship linking the end effector location (x_{tp}, y_{tp}) of the polymeric actuator to that of its corresponding joint angles subtended to their preceding one is given by Eqs. (5.1) and (5.2).

$$x_{tp} = l_1 \cos(\theta_1) + l_2 \cos(\theta_1 + \theta_2) + \cdots + l_n \cos(\theta_1 + \theta_2 + \cdots \theta_n) \quad (5.1)$$

$$y_{tp} = l_1 \sin(\theta_1) + l_2 \sin(\theta_1 + \theta_2) + \cdots + l_n \sin(\theta_1 + \theta_2 + \cdots \theta_n) \quad (5.2)$$

The polymer outline is superimposed on the approximated serial chain as shown in Figure 5.2, which allows a better visualization of the deduced model. $l_1, l_2, \ldots l_n$ represent the individual link lengths that assume angles $\theta_1, \theta_2, \ldots \theta_n$ with its each preceding link element.

In order to control the movement of the multi-body approximate of the polymer, IK methods are required that act towards reduction in the number of factors driving the model. IK presents an alternative approach of representing the complete set of joint parameters in terms of its desired end effector positions in a condensed Euclidian two-dimensional space. In order to solve the IK problem, it is mandatory that one must find the appropriate settings for the joint angles that ensure the resulting configuration of the multi-body to place each end effector at its targeted position. There are a number of available techniques originating from the realm of robotics that aim at resolving the IK problem. Popular approaches include the Jacobian pseudo-inverse and transpose approach [59,60], damped least squares (DLS) methods [61], neural network approaches [62] and cyclic coordinate descent (CCD) [63]. In the context of the present discussion that focusses on the application of IK procedures to animate soft actuator behaviour, an IK technique that can offer real-time animations of the entire multi-body by

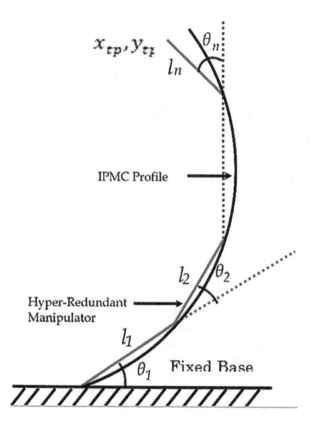

FIGURE 5.2 Actuator strip with the superimposed hyper-redundant model.

simultaneously tracking the end effector as well as conforming to the actuator trace is the one that is anticipated.

5.3.3 The Jacobian Transpose and Pseudo-Inverse Solution to Inverse Kinematics

The tip location of the modelled multi-body comprising of k discrete links is represented in terms of scalars $\theta_1, \theta_2, \ldots, \theta_n$ when fitted to an IK solver is poised to provide the complete actuator configuration in space. The forward kinematic problem described in Eqs. (5.1) and (5.2) may be extended to yield the forward dynamic formalism that represents the end effector velocities of each segment in terms of its corresponding joint velocities as shown in Eq. (5.3).

$$\left[\dot{x}_1\dot{y}_1, \dot{x}_2\dot{y}_2, \ldots, \dot{x}_{tp}\dot{y}_{tp}\right]^T = J(\theta)\left[\dot{\theta}_1, \dot{\theta}_2, \dot{\theta}_3, \ldots, \dot{\theta}_n\right]^T \quad (5.3)$$

The matrix $J \in \mathfrak{N}^{2k \times n}$ is known as the velocity Jacobian that relates the task space positional derivatives to those of the joint space. In case of the approximated

Time-Dependent Bending of IPMC Actuators

planar manipulator model, J is known as the manipulator Jacobian, which is expressed as a partial derivative of the two-dimensional Cartesian end effector in terms of its joint space components as shown in Eq. (5.4).

$$J = \begin{bmatrix} \frac{\partial x_1}{\partial \theta_1} & \frac{\partial x_2}{\partial \theta_2} & \cdots & \cdots & \frac{\partial x_{tp}}{\partial \theta_n} \\ \frac{\partial y_1}{\partial \theta_1} & \frac{\partial y_2}{\partial \theta_2} & \cdots & \cdots & \frac{\partial y_{tp}}{\partial \theta_n} \end{bmatrix} \quad (5.4)$$

In order to deduce the complete actuator trajectory $Q = [\ \theta_1,\ \theta_2,\ ...,\ \theta_n\ ]$ from its tracked tip locations $s \in \mathfrak{N}^{2k}$, the solution to the inverse problem may be viewed as an iterative procedure involving the pseudo-inverse of non-square Jacobian J (represented as $J^\dagger$), which is written the form of Eq. (5.5).

$$\dot{Q} = J^\dagger \dot{s} \quad (5.5)$$

This forms an incremental method of solving the overall IK, by seeking to reduce errors between the present and targeted end effectors $\vec{s}$ and $\vec{t}$, respectively. In order to do the same, an updation in the value of joint variable represented as for ΔQ its corresponding updation of Cartesian coordinate is performed. The computational burden associated with the calculation of Jacobian inverse may be circumvented by simply calculating its transpose, which is represented in the form of Eq. (5.6).

$$\Delta Q \approx J^T \cdot \Delta(\vec{t} - \vec{s}) \quad (5.6)$$

Both the IK procedures described above, however, yield a generic solution without the intervention of joint-specific limits, yet the same are limited in the case of hyper-redundant architectures. While an increase in redundancy offers a better resolution of the IK solution towards emulating flexibility, it fails to trace the desired profile in a way performed by the actuator and also incites jerky motions near singularities. Thus, a different solver is anticipated that is both applicable in the present context towards simulating actuation patterns and, at the same time, relevant in terms of computational prospects in offering solutions within the response time of the actual actuator.

5.3.4 'TRACTRIX'-BASED SOLUTION TO INVERSE KINEMATICS

A novel method exploiting the classical curve 'Tractrix' [64] was proposed in order to compute the IK for such serially linked multi-body systems. The tail location of a single mobile link was determined based on its desired head position. If the manipulator link is aligned to the vertical, and its head is slated to follow a straight path along the horizontal, the trajectory traversed by the manipulator tail is termed 'Tractrix'. The utility of employing the Tractrix-based IK [65] over Jacobian pseudo-inverse, backbone curve or modal approaches lies in the fact that it allows higher inertial weights to be assigned at the manipulator base, thereby ensuring a greater similarity of its traversed profile to the natural transition of approximated actuator joints [66]. The novelty of the approach lies in the proposition of an operative methodology that is capable of demarcating the marked natural shape variations of soft actuators without assigning superficial kinematic constraints to define their performance in response to changes in experimental media. It is noteworthy that the model being solely dependent on kinematic aspects and estranged from factors driving chemoelectrical variations within the polymer is expected to be successful in capturing the intrinsic diversity of its profiles.

The inverse kinematic problem of a flexible manipulator with its links designated as $l_1, l_2, l_3 \ldots, l_{n-1}, l_n$ while its corresponding joints being $j_1, j_2, j_3, \ldots, j_{n-1}, j_n$ in its order of assignment from its mounted base can be resolved iteratively by using the 'Tractrix'-based IK methodology described in Algorithm 5.1 [66]. Link l_1 and l_n are designated as the grounded link and the end effector of the flexible manipulator, respectively. The target for solving the IK is ascertained, based upon the polymer tip location, and subsequently provided as the head location of link l_n, while the tail location of the said link is calculated by feeding it to the devised routine. The calculated link tail then acts as the head for the penultimate link represented as l_{n-1} whose tail is subsequently calculated. The process is continual, until the tail location of the first link l_1 is found out.

Algorithm 5.1

1: Define vector $S = X_p - X_h$ connecting the current head location X_h and the destination head location X_p.
2: Define vector $T = X - X_h$ between $X = (x, y, z)^T$ which represents the tail of the link lying on the curve 'Tractrix' and the current head location X_h.
3: Assign a new coordinate system $\{r\}$ aligned to the X-axis and directed along S. $\hat{X}_r = S / |S|$.
4: Define the Z-axis as $Z_r = SXT/|SXT|$ orthogonal to both X_r and Y_r.
5: Define the rotation matrix $_r^0[R] = \begin{bmatrix} \hat{X}_r & \hat{Z}_r X \hat{X}_r & \hat{Z}_r \end{bmatrix}$ which relates the new coordinate frame to that of the Cartesian.
6: The y-coordinate of the link tail which lies on the curve is calculated as $y = \hat{Y}_r \cdot T$, while the parameter p can be obtained by using standard trigonometric transformations, which is derived as $p = L \cdot \mathrm{sech}^{-1}(y/L) \pm |S|$.
7: From p, the x- and y-coordinates of the point in the reference coordinate frame are obtained by and $y_r = L \mathrm{sech}(p/L)$, respectively.
8: The Cartesian coordinates with respect to the global reference frame $\{U\}$ are calculated by $(x, y, z)^T = X_h + _r^0[R](x_r, y_r, 0)^T$.

Time-Dependent Bending of IPMC Actuators

The various steps involved in the computation of the IK solution with the help of Tractrix-based IK algorithm in the global reference frame are pictorially represented in Figure 5.3. The described IK solver had originated in order to resolve redundancy in serial multi-body mobile platforms. However, in the context of a manipulator, the base being fixed mandates the displacement of the tail location of the link l_1 be subtracted from the position of all the links, which results in the joint motion to be constrained to execute solely rotary movements. This action results in a positional error between the end effector of the manipulator and its designated target location, which is minimized by executing the said algorithm iteratively. In this context, it is worth mentionable that the notion of 'Tractrix' emerged from the perspective of an object beginning its motion with a vertical offset and being dragged along a horizontal line along the X-axis. The described Tractrix algorithm extended the resulting motion in three-dimensional Cartesian space, for which an arbitrary position of the end effector was seen to drive the motion of all other links using the Tractrix solution. Once the end effector positions are known, the angular alignments of the manipulator are computed by resorting to simple vector algebra, thereby creating an effective solution to the incumbent redundancy. The algorithm was validated on an eight-link hyper-redundant mobile prototype, in addition to performing simulations of a moving snake and tying knots with a rope as examples of its potential applications [67]. The animated visuals were found to be more realistic since the displacement of the links diminished gradually from head to tail.

As a demonstration, the performance of the Tractrix-based IK solver has been evaluated by employing it to resolve the redundancy of a 20-link serial

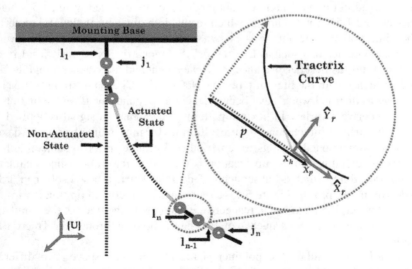

FIGURE 5.3 The motion of the actuator along with the approximated model with various frames of reference.

manipulator by making it follow two different ellipses. The dimensions of both the ellipses, one bearing its major axis aligned to the Y-axis, while the other having its major axis aligned to the X-axis, have been chosen to ensure that the motion remains within the manipulator workspace. The results of both the simulations are depicted in Figure 5.4a and b, respectively. In each set of depictions, four distinct instances in which the simulated manipulator using Tractrix is able to reach the designated trajectory are portrayed, represented in black.

From the simulation results of a hyper-redundant manipulator model using the 'Tractrix'-based IK solver, it is worth noticeable that the motion along the manipulator dies down from the leading head to the final tail, which qualifies its use for simulating motion for soft actuation elements. In addition to postural familiarity observed with that of soft actuators, the computational time required for performing the simulating motions for a 20-DoF model is considerably less due the involvement of simple linear algebraic techniques and hyperbolic functions over complex trigonometric calculations. This property also resonates well with the real-time response of such types of actuators while imitating their complete shape. The IK solver allows higher inertial weights to be assigned at the manipulator base in comparison with its distal end effector, thereby resulting in a greater similarity between the traversed profile of the simulation and the natural transition of approximated soft actuator joints.

5.3.4.1 Experimental Validation by Simulating Doped IPMC Actuators

In order to validate the action of the Tractrix IK solver, the same is used to simulate the actuation patterns of an electroactive polymer known as ionic polymer–metal composite (IPMC) when subjected to the voltage range of 0–5 V. The actuator had been soaked in three different dopant media, namely distilled water, 1.5 N NaCl solution and 1.5 N LiCl solution, which results in a diversity in flexural patterns. The ionic polymer membrane is known to absorb water molecules and several other cationic dopants such as Li^+, Na^+ or K^+. Among the listed varieties, Li^+ ion demonstrates the best performance in displacement and force generation [68–70]. In congruence with the literature review, NaCl and LiCl each of strength 1.5 N in addition to distilled water have been selected as dopants for IPMC actuation in order to derive a variety in flexural patterns. The actuation signal is applied on an IPMC strip of dimensions 40 mm × 10 mm × 0.2 mm, which has been soaked in the aforementioned dopants for a period of 5 minutes prior to subjecting it to electrical actuation, and its tip location is tracked in real time using a camera-based procedure, which is subsequently fed to the Tractrix-based solver in order to derive its full profile. Figure 5.5a–c represents the recorded tip trajectories for distilled water, LiCl and NaCl, respectively, along with the trace of the simulated model generated as a consequence of passing the tip data through the Tractrix IK routine.

In order to simulate the polymer shape in Cartesian space, two different approaches have been explored. In the first, the IK routine was executed only once, whereas in the other, the procedure was allowed to iterate until the tip of the flexible manipulator reached the targeted end effector location with acceptable

Time-Dependent Bending of IPMC Actuators

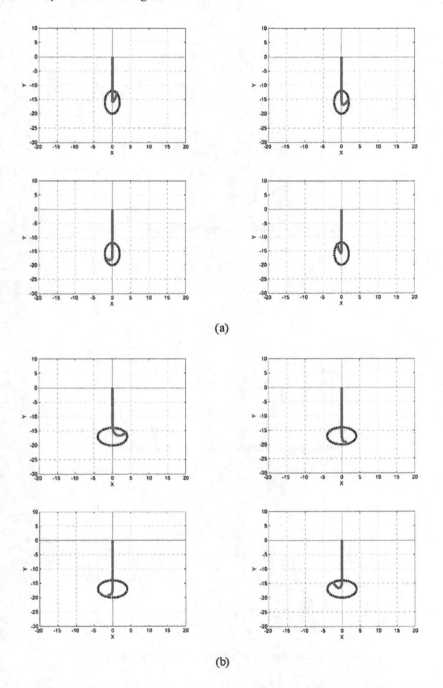

FIGURE 5.4 The simulated serial manipulator traversing an ellipse aligned to the (a) Y-axis and (b) X-axis.

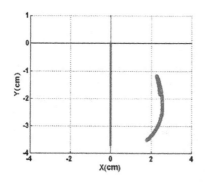 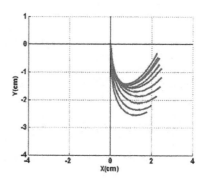

(a) Response of IPMC soaked in distilled water Solution

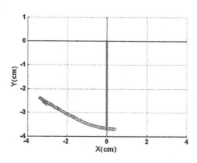 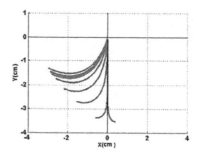

(b) Response of IPMC soaked in 1.5 N LiCl Solution

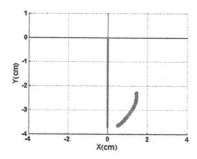 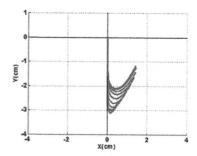

(c) Response of IPMC soaked in 1.5 N NaCl Solution

FIGURE 5.5 IPMC tip track data fitted to the 'Tractrix'-based IK routine. (a) Response of IPMC soaked in distilled water solution; (b) response of IPMC soaked in 1.5 N LiCl solution; (c) Response of IPMC soaked in 1.5 N NaCl solution.

TABLE 5.2
The Calculated Joint Positional Error for the IPMC Strip Doped in Various Solutions Obtained from Both the Modelling Routines

	Normalized Joint Position Error (mm)					
	Single Iteration			Multiple Iterations		
Link Number	Distilled Water	LiCl	NaCl	Distilled Water	LiCl	NaCl
1	0.0029	0.0601	0	0.1519	0.0766	0.0073
2	0	0.1189	0.0399	0.3375	0.1565	0.0810
3	0.0154	0.1489	0.1345	0.5287	0.2256	0.1918
4	0.0291	0.1350	0.1958	0.6618	0.2753	0.2849
5	0.0682	0.1200	0.2324	0.7923	0.3378	0.3714
6	0.1161	0.0749	0.2166	0.8937	0.3899	0.4314
7	0.1712	0.0294	0.1515	0.9600	0.4448	0.4703
8	0.2191	0.0271	0.0840	0.9953	0.5191	0.5259
9	0.2710	0.0309	0.0327	1.0000	0.5994	0.5824
10	0.3544	0.0276	0.1024	0.9419	0.6862	0.6518
11	0.4094	0	0.1548	0.8847	0.7774	0.7461
12	0.4732	0.0578	0.2006	0.7880	0.8751	0.8333
13	0.5026	0.1524	0.2058	0.6884	0.9595	0.9108
14	0.4975	0.2462	0.1417	0.5912	1.0000	0.9817
15	0.4646	0.3622	0.0584	0.4836	0.9879	1.0000
16	0.4084	0.4843	0.1383	0.3778	0.8883	0.9185
17	0.3869	0.6203	0.3409	0.2844	0.7209	0.7774
18	0.5139	0.7746	0.5530	0.2040	0.5163	0.5457
19	0.7193	0.8968	0.7720	0.0975	0.2690	0.2757
20	1.0000	1.0000	1.0000	0	0	0
Mean	**0.3312**	**0.2684**	**0.2378**	**0.5831**	**0.5353**	**0.5294**

levels of accuracy. The postural error represented as the deviation between the 20 joints of the simulated manipulator and that of the tracked IPMC segments was calculated and is listed in Table 5.2.

Analysing the data in the table, it may be safely inferred that the simulation of IPMC flexural patterns in the 1.5 N LiCl solution through the Tractrix routine is least efficient in terms of postural similarity between the simulated and actual trajectories since it yields the highest normalized mean joint positional error. The actual IPMC trace that has been tracked through the vision-based procedure is portrayed in Figures 5.6a, 5.7a and 5.8a, along with its approximated profile rendered by the modelling algorithm for distilled water, LiCl and NaCl depicted in Figures 5.6b and c, 5.7b and c, and 5.8b and c, respectively. A closer observation of the results reveals the existence of a significant trade-off between the capability of the Tractrix algorithm to track the polymer and its resulting cost of computation. Under the condition in which the routine ran for

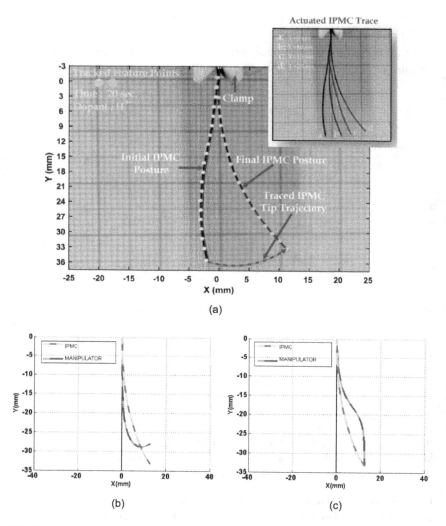

FIGURE 5.6 (a) Approximated IPMC manipulator soaked in distilled water; (b) single iteration; (c) multiple iterations.

a single iteration, it is apparent that despite the curve maintaining a bending profile similar to the actuated IPMC trace, the desired location was far from being reached (represented by the highest joint error at end effector). On the other hand, the quest to reach a given target (represented by zero normalized error at end effector) results in the profile being deviated from that of the traced polymer for the second case where the routine executes multiple iterations. The cases draw motivation from Tractrix being applied to a mobile robot platform and that to a serial manipulator, respectively. From the plot, it is also evident that the joint angles of each individual link of the manipulator model draw a smooth

Time-Dependent Bending of IPMC Actuators 111

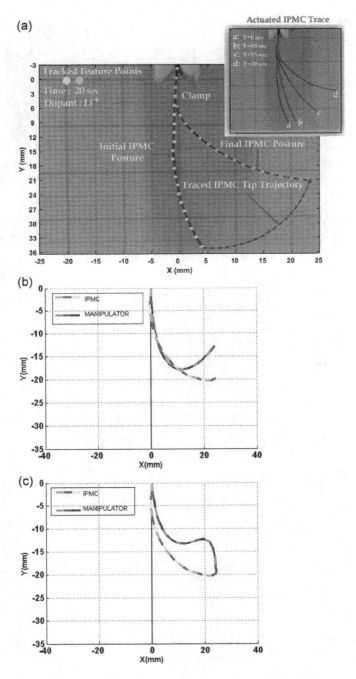

FIGURE 5.7 (a) Approximated IPMC manipulator soaked in 1.5 N LiCl solution; (b) single iteration; (c) multiple iterations.

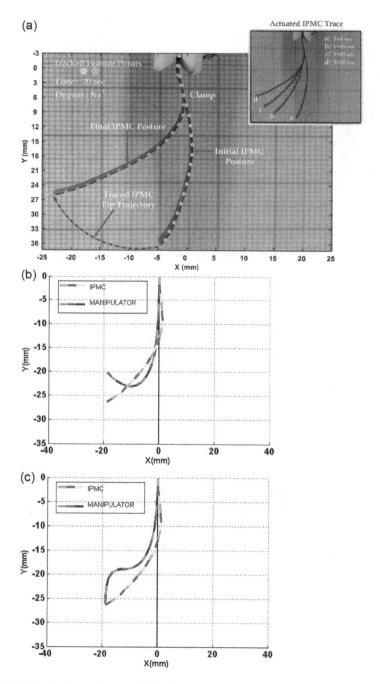

FIGURE 5.8 (a) Approximated IPMC manipulator soaked in 1.5 N NaCl solution; (b) single iteration; (c) multiple iterations.

transition while traversing from one link to its adjoining link, which yields a match with the physical structure of IPMC in space.

From the labelled Figures 5.6 to 5.8, it is imperative that polymer flexion under the influence of distilled water results in a subtle temporal graduation of joint angle values. A significantly large initial angular variation at the actuator basal link paves the way for an eventually high tip displacement in the case of polymers soaked in LiCl solutions. In stark contrast, link angles at the distal portion of the actuator doped with NaCl register higher variations; hence, the overall displacement of the polymer tip is rendered negligible. The joint angles for the above scenario are noted to be maximum as well as uniform over all instances, endorsing the fact that infinitesimal proximal angular variations are compensated for by a corresponding high distal deviation. The 'Tractrix'-based IK routine discussed in Section 3.4 has been validated successfully by resolving the hyper-redundancy of a flexible electroactive polymer such as IPMC. The discussed IK solver by its inherent nature is seen to yield a natural flexural pattern with its individual joint angles cumulatively increasing from the basal link to the manipulator end effector. The procedure ensures an effective emulation of the IPMC actuator profile in space without putting additional joint boundaries or angular constraints. In order to put things in perspective by demonstrating a practical application of the modelling scheme, the developed modelling algorithm is used to predict the workspace clearance generated by such an actuator while articulating small objects of various shapes, which has been described in the subsequent sections.

5.4 DESIGN AND DEVELOPMENT OF COMPLIANT SOFT ACTUATED GRIPPERS: AN APPLICATION OF HYPER-REDUNDANT KINEMATIC MODELLING

The Tractrix-based IPMC shape differentiation approach has established the dissimilarities in polymer profile in distilled water and LiCl solution (the two cases represent the extremes of generated deflection across various dopants) on quantifiable kinematic terms. The modelling algorithm devised in the previous section is used to study the effect of ionic liquids on workspace characteristics of an IPMC actuated gripper. The simulation is destined to play a pivotal role in determining the size and profile of the object that could be negotiated by such a gripping device. Two distinct designs, namely a two-jaw active IPMC gripper and a PDMS (polydimethylsiloxane) jaw IPMC gripper, have been analysed in the present study. The grippers are investigated for their physical characteristics in the two solutions (distilled water and 1.5 N LiCl), with the aim of exploring a broader aspect of dopant-specific gripper assembly that is capable of offering an option to select a particular configuration based on the prior knowledge of the gripper jaw workspace. Although IPMC actuated gripping devices have been investigated for quite a while, the evaluation of the workspace in quantifiable terms has been found to lack in all prior investigations. A complete list of IPMC actuated gripping devices across their chronological evolution is presented in Table 5.1. Traversing through the list, it is apparent that among all gripper configurations, a parallel jaw gripper is the most common configuration that has been devised over the years. Its marked simplicity in design aspects as well as its robustness to accomplish a task has made it a widely accepted scheme. Also, it is noticeable that apart from a double active-jaw configuration, a passive jaw gripper is popular among investigators, despite its low load-carrying capabilities due to the vital force information obtained from the intrinsic structural design [71].

The workspace of both these configurations is modelled using the Tractrix IK routine [72] by subjecting both the IPMC and PDMS tips to Algorithm 5.1. The generated workspace for both dopants is shown in Figures 5.9a–d and 5.10a–d, respectively. From the figures, it is evident that the simulation routine is able to capture the characteristics of the gripper assembly. Significant inferences such as the nature of workspace and payload capacity of the gripper may be derived from the results. It is seen that in both active and passive jaw gripper configurations, LiCl-doped samples portray a larger workspace. It is also notable that since the same portrays a characteristic curvilinearity in flexural patterns in contrast to the hydrated polymer samples, LiCl grippers may attract more relevance in handling spherical payloads over the ones possessing sharp edges such as a cube which may be adroit for a gripper soaked in distilled water. Due to greater simulated deflections, LiCl grippers are expected to carry heavier objects in comparison with their distilled water counterparts which are apt for generating a point contact on the handled object [73]. Thus, the modelling methodology provides a deeper insight into the specific characteristics of various gripper designs prior to their real-time operation.

5.5 CONCLUSIONS AND FUTURE PROSPECTS

This chapter has sought to address the domain of mechanics-based modelling approaches that have emerged as a viable alternative to simulate motions of soft-polymer actuators. At first, it provides a general overview of the terminology of 'soft robotic' materials by discussing various material alternatives available in the present context, following which common modelling methodologies and their associated challenges are discussed from the traditional viewpoint of electromechanical characterization. Finally, the domain of inverse kinematics-based simulation has been presented. Traditional IK approaches involving the Jacobian pseudo-inverse or transpose have been illustrated with their associated incapabilities. A new IK technique based on the classical curve 'Tractrix' that generates motion similar to natural actuation patterns is presented as an alternative to the prior schemes. The simulated motion is seen to bear strong resemblance to the flexural motions generated upon electrical actuations. The modelling algorithm is finally validated by discussing a practical application of robotics in the domain of soft robotic gripper development, which is seen to augment conventional approaches by offering a priori information regarding the gripper workspace before engaging in actual manipulation. The entire scheme shares prospects of further development, where the developed procedure may be used in an integrated micro-manipulation assembly involving soft robotic grippers where the knowledge of the simulation would be incorporated within its design.

Time-Dependent Bending of IPMC Actuators

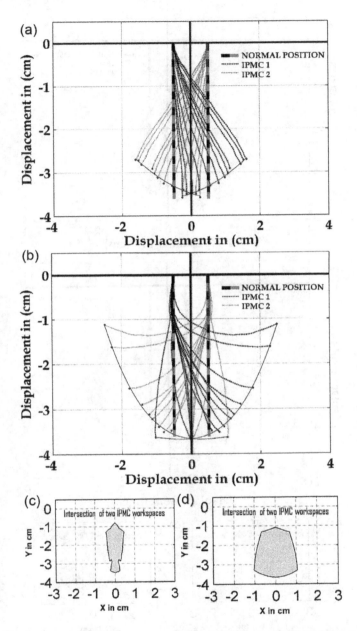

FIGURE 5.9 Profile and workspace of a two-jaw IPMC gripper in distilled water (a–c) and 1.5 N LiCl solution (b–d).

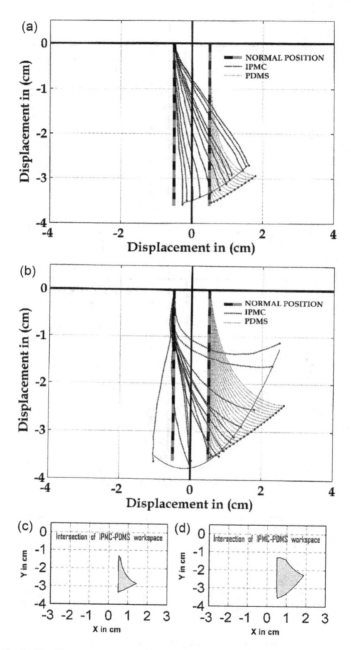

FIGURE 5.10 Profile and workspace of a passive PDMS jaw IPMC gripper in distilled water (a and c) and 1.5 N LiCl solution (b and d).

ACKNOWLEDGEMENTS

The authors would like to acknowledge all the sources from whom the information contained in this presentation was collated. Specifically, they acknowledge the Indo-US Centre for Research Excellence on Fabrionics at IIEST Shibpur, and Chemistry and Biomimetics Laboratory, CSIR-CMERI, Durgapur, for providing motivation and financial support.

In the next chapter, the reader will find the methods of material selection for soft engineering applications; also, in Figure 6.1, the research methodology graph for the characterization of IPMC, tele-operation, soft microgripper design, etc., is explained. For beginner and for new ideas can be dogged from that chapter.

REFERENCES

[1] Wang L, Iida F. Deformation in soft-matter robotics: A categorization and quantitative characterization. *IEEE Robotics & Automation Magazine*. 2015 Sep; 22(3):125–39.

[2] Trivedi D, Rahn CD, Kier WM, Walker ID. Soft robotics: Biological inspiration, state of the art, and future research. *Applied Bionics and Biomechanics*. 2008 Dec 16; 5(3):99–117.

[3] Albu-Schaffer A, Eiberger O, Grebenstein M, Haddadin S, Ott C, Wimbock T, Wolf S, Hirzinger G. Soft robotics. *IEEE Robotics & Automation Magazine*. 2008 Sep; 15(3):20–30.

[4] Shahinpoor M, Kim KJ. Ionic polymer–metal composites: III. Modeling and simulation as biomimetic sensors, actuators, transducers, and artificial muscles. *Smart Materials and Structures*. 2004 Oct 6; 13(6):1362.

[5] Jo C, Pugal D, Oh IK, Kim KJ, Asaka K. Recent advances in ionic polymer–metal composite actuators and their modeling and applications. *Progress in Polymer Science*. 2013 Jul 31; 38(7):1037–66.

[6] Otero TF, Martinez JG, Arias-Pardilla J. Biomimetic electrochemistry from conducting polymers. A review: Artificial muscles, smart membranes, smart drug delivery and computer/neuron interfaces. *Electrochimica Acta*. 2012 Dec 1; 84:112–28.

[7] Kim KJ, Tadokoro S. *Electroactive Polymers for Robotic Applications. Artificial Muscles and Sensors* (291 p.). London, United Kingdom: Springer. 2007.

[8] Bar-Cohen Y. Electroactive polymers as artificial muscles-capabilities, potentials and challenges. Handbook on Biomimetics. 2000 Apr:8.

[9] Shahinpoor M, Kim KJ. Ionic polymer–metal composites: IV. Industrial and medical applications. *Smart Materials and Structures*. 2004 Dec 23; 14(1):197.

[10] Kakugo A, Shikinaka K, Gong JP, Osada Y. Gel machines constructed from chemically cross-linked actins and myosins. *Polymer*. 2005 Aug 23; 46(18):7759–70.

[11] Soong RK, Bachand GD, Neves HP, Olkhovets AG, Craighead HG, Montemagno CD. Powering an inorganic nanodevice with a biomolecular motor. *Science*. 2000 Nov 24; 290(5496):1555–8.

[12] Morishima K, Tanaka Y, Sato K, Ebara M, Shimizu T, Yamato M, Kikuchi A, Okano T, Kitamori T. Bio actuated microsystem using cultured cardiomyocytes. Proceedings of the MicroTotal Analysis Systems, Squaw Valley, CA. 2003:1125–8.

[13] Xi J, Schmidt JJ, Montemagno CD. Self-assembled microdevices driven by muscle. *Nature Materials*. 2005 Feb 1; 4(2):180–4.

[14] Tabiryan N, Serak S, Dai XM, Bunning T. Polymer film with optically controlled form and actuation. *Optics Express*. 2005 Sep 19; 13(19):7442–8.

[15] Yamada M, Kondo M, Mamiya JI, Yu Y, Kinoshita M, Barrett CJ, Ikeda T. Photomobile polymer materials: Towards light-driven plastic motors. *Angewandte Chemie International Edition*. 2008 Jun 23; 47(27):4986–8.

[16] Pelrine RE, Kornbluh RD, Joseph JP. Electrostriction of polymer dielectrics with compliant electrodes as a means of actuation. *Sensors and Actuators A: Physical*. 1998 Jan 1; 64(1):77–85.

[17] Pelrine R, Kornbluh R, Pei Q, Joseph J. High-speed electrically actuated elastomers with strain greater than 100%. *Science*. 2000 Feb 4; 287(5454):836–9.

[18] Pelrine R, Kornbluh R, Kofod G. High-strain actuator materials based on dielectric elastomers. *Advanced Materials*. 2000 Aug 1; 12(16):1223–5.

[19] Kurita Y, Ueda T, Kasazaki T, Hirai T (inventors); Nitta Corporation (assignee). Polyurethane Elastomer Actuator. United States patent US 5,977,685. 1999 Nov 2.

[20] Zhang QM, Bharti V, Zhao X. Giant electrostriction and relaxor ferroelectric behavior in electron-irradiated poly (vinylidene fluoride-trifluoroethylene) copolymer. *Science*. 1998 Jun 26; 280(5372):2101–4.

[21] Lehmann W, Hartmann L, Kremer F, Stein P, Finkelmann H, Kruth H, Diele S. Direct and inverse electromechanical effect in ferroelectric liquid crystalline elastomers. *Journal of Applied Physics*. 1999 Aug 1; 86(3):1647–52.

[22] Lehmann W, Skupin H, Tolksdorf C, Gebhard E, Zentel R, Krüger P, Lösche M, Kremer F. Giant lateral electrostriction in ferroelectric liquid-crystalline elastomers. *Nature*. 2001 Mar 22; 410(6827):447–50.

[23] Baughman RH, Cui C, Zakhidov AA, Iqbal Z, Barisci JN, Spinks GM, Wallace GG, Mazzoldi A, De Rossi D, Rinzler AG, Jaschinski O. Carbon nanotube actuators. *Science*. 1999 May 21; 284(5418):1340–4.

[24] Hughes M, Spinks GM. Multiwalled carbon nanotube actuators. *Advanced Materials*. 2005 Feb 23; 17(4):443–6.

[25] Aliev AE, Oh J, Kozlov ME, Kuznetsov AA, Fang S, Fonseca AF, Ovalle R, Lima MD, Haque MH, Gartstein YN, Zhang M. Giant-stroke, superelastic carbon nanotube aerogel muscles. *Science*. 2009 Mar 20; 323(5921):1575–8.

[26] Lima MD, Li N, De Andrade MJ, Fang S, Oh J, Spinks GM, Kozlov ME, Haines CS, Suh D, Foroughi J, Kim SJ. Electrically, chemically, and photonically powered torsional and tensile actuation of hybrid carbon nanotube yarn muscles. *Science*. 2012 Nov 16; 338(6109):928–32.

[27] Phillips CL, Jankowski E, Krishnatreya BJ, Edmond KV, Sacanna S, Grier DG, Pine DJ, Glotzer SC. Digital colloids: Reconfigurable clusters as high information density elements. *Soft Matter*. 2014; 10(38):7468–79.

[28] Okuzaki H, Kunugi T. Electrically induced contraction of polypyrrole film in ambient air. *Journal of Polymer Science Part B: Polymer Physics*. 1998 Jul 15; 36(9):1591–4.

[29] Okuzaki H, Suzuki H, Ito T. Electromechanical properties of poly (3, 4-ethylenedioxythiophene)/poly (4-styrene sulfonate) films. *The Journal of Physical Chemistry B*. 2009 Jul 29; 113(33):11378–83.

[30] Ma M, Guo L, Anderson DG, Langer R. Bio-inspired polymer composite actuator and generator driven by water gradients. *Science*. 2013 Jan 11; 339(6116):186–9.

[31] Okuzaki H, Kunugi T. Adsorption-induced bending of polypyrrole films and its application to a chemomechanical rotor. *Journal of Polymer Science Part B: Polymer Physics*. 1996 Jul 30; 34(10):1747–9.

[32] Okuzaki H, Funasaka K. Electromechanical properties of a humido-sensitive conducting polymer film. *Macromolecules*. 2000 Oct 31; 33(22):8307–11.

[33] Smela E, Inganas O, Lundstrom I. Controlled folding of micrometer-size structures. *Science*. 1995 Jun 23; 268(5218):1735.
[34] Kaneto K, Kaneko M, Min Y, MacDiarmid AG. "Artificial muscle": Electromechanical actuators using polyaniline films. *Synthetic Metals*. 1995 Apr 1; 71(1):2211–2.
[35] Takashima W, Kaneko M, Kaneto K, MacDiarmid AG. The electrochemical actuator using electrochemically-deposited poly-aniline film. *Synthetic Metals*. 1995 Apr 1; 71(1):2265–6.
[36] Lu W, Fadeev AG, Qi B, Smela E, Mattes BR, Ding J, Spinks GM, Mazurkiewicz J, Zhou D, Wallace GG, MacFarlane DR. Use of ionic liquids for π-conjugated polymer electrochemical devices. *Science*. 2002 Aug 9; 297(5583):983–7.
[37] Hara S, Zama T, Takashima W, Kaneto K. TFSI-doped polypyrrole actuator with 26% strain. *Journal of Materials Chemistry*. 2004; 14(10):1516–7.
[38] Kwon GH, Park JY, Kim JY, Frisk ML, Beebe DJ, Lee SH. Biomimetic soft multifunctional miniature aquabots. *Small*. 2008 Dec 1; 4(12):2148–53.
[39] Liu Z, Calvert P. Multilayer hydrogels as muscle-like actuators. *Advanced Materials*. 2000 Feb 1; 12(4):288–91.
[40] Schexnailder P, Schmidt G. Nanocomposite polymer hydrogels. *Colloid and Polymer Science*. 2009 Jan 1; 287(1):1–1.
[41] Sreekumar M, Nagarajan T, Singaperumal M, Zoppi M, Molfino R. Critical review of current trends in shape memory alloy actuators for intelligent robots. *Industrial Robot: An International Journal*. 2007 Jun 26; 34(4):285–94.
[42] Ikuta K. Micro/miniature shape memory alloy actuator. In Robotics and Automation, 1990. Proceedings 1990 IEEE International Conference on. 1990 May 13 (pp. 2156–2161). IEEE.
[43] Stirling L, Yu CH, Miller J, Hawkes E, Wood R, Goldfield E, Nagpal R. Applicability of shape memory alloy wire for an active, soft orthotic. *Journal of Materials Engineering and Performance*. 2011 Jul 1; 20(4–5):658–62.
[44] Paik JK, Hawkes E, Wood RJ. A novel low-profile shape memory alloy torsional actuator. *Smart Materials and Structures*. 2010 Nov 17; 19(12):125014.
[45] Crawley EF, De Luis J. Use of piezoelectric actuators as elements of intelligent structures. *AIAA Journal*. 1987 Oct; 25(10):1373–85.
[46] Caldwell DG, Medrano-Cerda GA, Goodwin M. Control of pneumatic muscle actuators. *IEEE Control Systems*. 1995 Feb; 15(1):40–8.
[47] Katchalsky A. Rapid swelling and deswelling of reversible gels of polymeric acids by ionization. *Cellular and Molecular Life Sciences*. 1949 Aug 29; 5(8):319–20.
[48] Kenaley GL, Cutkosky MR. Electrorheological fluid-based robotic fingers with tactile sensing. In Robotics and Automation, 1989. Proceedings 1989 IEEE International Conference on. 1989 May 14 (pp. 132–136). IEEE.
[49] Carlson JD, Catanzarite DM, St. Clair KA. Commercial magneto-rheological fluid devices. *International Journal of Modern Physics B*. 1996 Oct 30; 10(23n24):2857–65.
[50] Scrosati B (ed). *Applications of Electroactive Polymers*. London: Chapman & Hall. 1993 Jul.
[51] Kim KJ, Tadokoro S. Electroactive polymers for robotic applications. *Artificial Muscles and Sensors* (291 p.). London, United Kingdom: Springer. 2007.
[52] Bar-Cohen Y. Electroactive polymers: Current capabilities and challenges. In SPIE's 9th Annual International Symposium on Smart Structures and Materials. 2002 Jul 10 (pp. 1–7). International Society for Optics and Photonics.
[53] Pei Q, Inganäs O. Electrochemical applications of the bending beam method. 1. Mass transport and volume changes in polypyrrole during redox. *The Journal of Physical Chemistry*. 1992 Dec; 96(25):10507–14.

[54] Pei Q, Inganaes O. Electrochemical applications of the bending beam method. 2. Electroshrinking and slow relaxation in polypyrrole. *The Journal of Physical Chemistry*. 1993 Jun; 97(22): 6034–41.

[55] Benslimane M, Gravesen P, West K, Skaarup S, Sommer-Larsen P. Performance of polymer-based actuators: The three-layer model. In 1999 Symposium on Smart Structures and Materials. 1999 May 28 (pp. 87–97). International Society for Optics and Photonics.

[56] Alici G, Mui B, Cook C. Bending modeling and its experimental verification for conducting polymer actuators dedicated to manipulation applications. *Sensors and Actuators A: Physical*. 2006 Feb 14; 126(2):396–404.

[57] Alici G. An effective modelling approach to estimate nonlinear bending behaviour of cantilever type conducting polymer actuators. *Sensors and Actuators B: Chemical*. 2009 Aug 18; 141(1):284–92.

[58] Mutlu R, Alici G, Li W. Electroactive polymers as soft robotic actuators: Electromechanical modeling and identification. In 2013 IEEE/ASME International Conference on Advanced Intelligent Mechatronics. 2013 Jul 9 (pp. 1096–1101). IEEE.

[59] Whitney DE. Resolved motion rate control of manipulators and human prostheses. *IEEE Transactions on Man-Machine Systems*. 1969 June; 10(2):47–53.

[60] Wolovich WA, Elliott H. A computational technique for inverse kinematics. In Decision and Control, 1984. The 23rd IEEE Conference on. 1984 Dec 12 (pp. 1359–1363). IEEE.

[61] Wampler CW. Manipulator inverse kinematic solutions based on vector formulations and damped least-squares methods. *IEEE Transactions on Systems, Man, and Cybernetics*. 1986 Jan; 16(1):93–101.

[62] Tevatia G, Schaal S. Inverse kinematics for humanoid robots. In Robotics and Automation, 2000. Proceedings. ICRA'00. IEEE International Conference on. 2000 (Vol. 1, pp. 294–99). IEEE.

[63] Wang LC, Chen CC. A combined optimization method for solving the inverse kinematics problems of mechanical manipulators. *IEEE Transactions on Robotics and Automation*. 1991 Aug; 7(4):489–99.

[64] Steinhaus H. *Mathematical Snapshots*. Courier Corporation. 2012 Jul 12.

[65] Sreenivasan S, Goel P, Ghosal A. A real-time algorithm for simulation of flexible objects and hyper-redundant manipulators. *Mechanism and Machine Theory*. 2010 Mar 31; 45(3):454–66.

[66] Chattaraj R, Khan S, Bhattacharya S, Bepari B, Chatterjee D, Bhaumik S. Shape estimation of IPMC actuators in ionic solutions using hyper redundant kinematic modeling. *Mechanism and Machine Theory*. 2016 Sep 30; 103:174–88.

[67] Ravi VC, Rakshit S, Ghosal A. Redundancy resolution using tractrix—simulations and experiments. *Journal of Mechanisms and Robotics*. 2010 Aug 1; 2(3):031013.

[68] Bhandari B, Lee GY, Ahn SH. A review on IPMC material as actuators and sensors: Fabrications, characteristics and applications. *International Journal of Precision Engineering and Manufacturing*. 2012 Jan 1; 13(1):141–63.

[69] Yeh CC, Shih WP. Effects of water content on the actuation performance of ionic polymer–metal composites. *Smart Materials and Structures*. 2010 Nov 11; 19(12):124007.

[70] Shahinpoor M, Kim KJ. The effect of surface-electrode resistance on the performance of ionic polymer-metal composite (IPMC) artificial muscles. *Smart Materials and Structures*. 2000 Aug; 9(4):543.

[71] Chattaraj R, Bhattacharya S, Bepari B, Bhaumik S. Design and control of two fingered compliant gripper for micro gripping. In Informatics, Electronics & Vision (ICIEV), 2014 International Conference on. 2014 May 23 (pp. 1–6). IEEE.

[72] Chattaraj R, Khan S, Bhattacharya S, Bepari B, Chatterjee D, Bhaumik S. Development of two jaw compliant gripper based on hyper-redundant approximation of IPMC actuators. *Sensors and Actuators A: Physical*. 2016 Nov 1; 251: 207–18.

[73] Chattaraj R, Bhattacharya S, Roy A, Mazumdar A, Bepari B, Bhaumik S. Gesture based control of IPMC actuated gripper. In Engineering and Computational Sciences (RAECS), 2014 Recent Advances in. 2014 Mar 6 (pp. 1–6). IEEE.

[74] Feng GH, Yen SC. Micromanipulation tool replaceable soft actuator with gripping force enhancing and output motion converting mechanisms. In 2015 Transducers-2015 18th International Conference on Solid-State Sensors, Actuators and Microsystems (TRANSDUCERS). 2015 Jun 21 (pp. 1877–80). IEEE.

[75] Bhattacharya S, Bepari B, Bhaumik S. Novel approach of IPMC actuated finger for micro-gripping. In Informatics, Electronics & Vision (ICIEV), 2015 International Conference on. 2015 Jun 15 (pp. 1–6). IEEE.

6 Selection of Elastomer for Compliant Robotic Gripper Harnessed with IPMC Actuator

Srijan Bhattacharya
RCC Institute of Information Technology

Bikash Bepari
Haldia Institute of Technology

Subhasis Bhaumik
Indian Institute of Engineering Science and Technology

CONTENTS

- 6.1 Introduction 124
- 6.2 Multi-Criteria Decision Making (MCDM) Problem Formulation 128
- 6.3 Technique for Order of Preference by Similarity to Ideal Solution 129
- 6.4 Complex Proportional Assessment 130
- 6.5 Multi-Objective Optimization on the Basis of Ratio Analysis 132
- 6.6 ELECTRE II 133
- 6.7 Determination of Entropy Weight 135
- 6.8 Spearman's Rank Correlation Coefficient 136
- 6.9 Criteria for Compliant Material Selection 136
  - 6.9.1 Hardness 136
  - 6.9.2 Density 137
  - 6.9.3 Tensile Strength 137
  - 6.9.4 Elongation at Break 137
  - 6.9.5 Cost 138
- 6.10 Compliant Materials 138
  - 6.10.1 Ethylene–Propylene Diene Monomer (EPDM) 138
  - 6.10.2 Ethylene–Vinyl Acetate (EVA) 138
  - 6.10.3 Ethylene–Propylene Monomer (EPM) 138
  - 6.10.4 Polydimethylsiloxane (PDMS) 139
  - 6.10.5 Polyurethane (PU) 139

DOI: 10.1201/9781003204664-6

        6.10.6 Ethylene–Propylene Terpolymer (EPT) 139
        6.10.7 Polyvinylidene Fluoride (PDVF) 139
6.11 Results 139
        6.11.1 Weight Determination by Entropy Method 140
        6.11.2 Solution by TOPSIS 141
        6.11.3 Solution by COPRAS 141
        6.11.4 Solution by MOORA 142
        6.11.5 Solution by ELECTRE II 143
6.12 Discussion 144
6.13 Conclusions 144
References 145

6.1 INTRODUCTION

Over the past couple of decades, multi-criteria decision making (MCDM) has pioneered itself as an imperative as well as inevitable tool for decision support systems which fundamentally necessitate multiple decision alternatives, to adjudge their suitability subjected to a finite set of attributes or criteria (Yue [1], Hwang and Yoon [2] and Zeleny [3]). MCDM techniques such as the technique for order of preference by similarity to ideal solution (TOPSIS) method (Milani et al. [4] and Caliskan et al. [5]), VIseKriterijumska Optimizacija I Kompromisno Resenje (VIKOR) method (Opricovic and Tzeng [6]; Caliskan et al. [5]), simple additive weighting (SAW) method (Memariani [7]), multi-objective optimization on the basis of ratio analysis (MOORA) (Brauers and Zavadskas [8]; Karande and Chakraborty [9]), preference selection index (PSI) (Maniya and Bhatt [10]), graph theory matrix approach (Rao and Padmanabhan [11]), complex proportional assessment (COPRAS) (Antucheviciene et al. [12]), elimination and choice expressing the reality method (ELECTRE) (Pang et al. [13]), evaluation of mixed data (EVAMIX) method (Chatterjee and Chakraborty [14]), EXPROM (Chatterjee and Chakraborty [15]), multi-attribute utility theory (MAUT) (Roth et al. [16]), quality function deployment (QFD) (Prasad and Chakraborty [17]), digital logic method (Manshadi et al. [18]), utility additive method (UTA) (Athawale et al. [19]), operational competitiveness rating analysis (OCRA) (Parkan and Wu [20]) and multi-criteria decision making based on ordinal data (MCDM-BOD) (Jahan et al. [21]) have effectively been employed to solve multi-criteria selection problems for different engineering applications. All the above-mentioned methods are bestowed with unique mathematical and logical foundations pertinent to the scope of applications.

The authors revealed during the literature survey that the assortment of compliant materials for robotic applications is narrow in nature. Nevertheless, the following literature review will unveil few materials that have been used for compliant materials gripping purpose. Reddy et al. [22] presented the design, development and validation of compliant miniature grippers comprising of parallel and angular jaw motions. The grippers were made up of spring steel and polydimethylsiloxane (PDMS) materials. In their study, grippers were employed to grip yeast balls and zebrafish eggs which were nearly 1 mm in diameter. Zubir et al. [23] developed a novel design of microgripper by amalgamating bias spring and flexure hinge concepts to poise it up for obtaining high-accuracy microgripping. The prototype compliant gripper was made from aluminum alloy (Al 7075T6) in wire EDM machine. Changa et al. [24] developed a prototype mesoscopic gripping system made of polyurethane (PU) along with actuator subsystem, and sensor subsystem. The operational accuracy of mechanism was around 5 μm with a gripping range up to 2 mm. The payload was observed as 0.5 g. It was observed that the PU gripper is less prone to failure due to cyclic loading at the compliant joints. Zhang et al. [25] designed and developed a SPCA-driven monolithic compliant microgripper (MCM) made of aluminum alloy (type: 7075T6), which was fabricated by wire EDM technology keeping in view of micro-assembly. Moreover, they conducted a series of experiments to validate the kinematic model and to establish control strategy during microgripping. They were able to clamp a glassy microtube of 150 μm diameter by their developed MCM. Voigt et al. [26] developed a lightweight robot capable of climbing in varied environments. Eventually, the situation demands for tribologically optimized materials at the juncture of the object being grasped and gripping device. Their goal was to study and screen different polymer materials which pose with substantial frictional, cohesive and stiffness attributes. Their study revealed that thin polymeric films substrated with rubber foamy materials provided excellent compliant properties. Bhattacharya et al. [27] developed an IPMC actuated compliant two-jaw gripper fabricated with a PDMS material. The gripper was found to manipulate up to 1.247 g of weight. Bahraminasab et al. [28] selected the best alternative from a set of materials for femoral component of total knee replacement with the help of VIKOR method. The best option was reached at, and sensitivity analysis of weights was taken up to reach exact results. The resulted ranking order revealed that porous and dense NiTi shape-memory alloys achieved first and second position, respectively. Alemi et al. [29] used the ELECTRE model to optimize the selection of artificial lift systems in petroleum industry and confirmed with a number of certain oil fields data. They reached a substantial agreement between ELECTRE model program final results and the field experimental results. Jee et al. [30] used decision making theory to calculate the weightage for each alternative. They also used TOPSIS method to evaluate the ranking for the respective alternatives considering all the parameters and respective weights. Rao et al. [31] selected the most favorable option for a design problem from a given set of alternatives with the help of multi-attribute decision making (MADM). They proposed a de novo approach to solve the problem by considering the weights as well as biased preferences of the decision maker to calculate the integrated weights. Moreover, the authors also used fuzzy logic method to change the qualitative criteria into quantitative criteria. Jahan et al. [32] focused on the importance of target values besides cost and benefit criteria while selecting optimal solution for replacing human tissue in biomedical engineering applications. They proposed a novel normalization technique, which is an extension of the TOPSIS method and objective weighting in accordance with the selection of materials. They cited four examples to confirm the accuracy of outcomes from the said method cases included to validate the accuracy of outcomes from the proposed model. Jahan et al. [33] proposed an aggregation technique for optimal decision making, wherein they used ranking orders of various alternatives obtained by various MCDM methods as input and the outputs as aggregation ranks, to help designers and engineers arrive at a consensus on materials selection for any particular application. Ipek et al. [34] tried to solve the materials selection problem by considering different attributes for automotive parts such as impact resistance, lightness, formability, corrosion resistance and low prices for bumpers; strength, formability, vibration absorption and low cost for flywheels; and strength, formability, corrosion resistance, biocompatibility and a small price for implants. Afterward, they made this method more capable with the help of expert system approach. After they considered the attributes, they found that polymeric materials such as PP (polypropylene), HDPE (high-density polyethylene) and PMMA [poly(methyl methacrylate)] for the bumpers; GFRPs (glass fiber-reinforced plastics) and CFRPs (carbon fiber-reinforced plastics) composites for high-speed running, and cast iron and steel for low speeds for the flywheels; and finally stainless steel and polymeric materials {such as PVC (polyvinyl chloride) and PE (polyethylene)} were found to be the best materials for automotive parts. Cicek et al. [35] proposed an integrated decision aid (IDEA), wherein suitable techniques match various problem statements based on six dimensions, namely the type of the decision problem, size of the problem, selection of the preference techniques by decision makers, decision makers' preference structure, necessity for the use of relative importance and lastly the nature of performance values. Moreover, the implementation procedure of the proposed IDEA for a material selection problem was demonstrated with the previously cited applications from the materials science literature. The method IDEA provided great advantages and encouragements to research fellows in lieu of preventing unnecessary time consumption, plausible misapplications and numerous challenges in regard to multi-criteria analysis of material selection problems. Anojkumar et al. [36] intended to portray the application of four MCDM methods for solving pipes material selection problem in sugar industry. The best alternative amid numerous materials was chosen using the four methods, namely FAHP-TOPSIS, FAHP-VIKOR, FAHP-ELECTRE and FAHP-PROMTHEE. The ranking performance of numerous MCDM methods was also adjudged among the others, and the effectiveness and flexibility of the VIKOR method were also explored.

There are numerous materials with diverse properties, and because this selection of the suitable material is very time-consuming, it is also complicated. Hence, it is important to find a systematic and effective approach to material selection problem. Five stainless steel grades such as J4, JSLAUS, J204Cu, 409M and 304 and seven evaluation criteria such as yield strength, ultimate tensile strength, percentage of elongation, hardness, cost, corrosion rate and wear rate were considered in the paper for finding the suitable material among the others. Kan et al. [37] tried to find the appropriate material for manufacturing the tool holder for hard milling operation using a decision model, including extended PROMETHEE II (EXPROM2) (preference ranking organization method for enrichment evaluation), TOPSIS (technique for order performance by similarity to ideal solution) and VIKOR (VIseKriterijumska Optimizacija I Kompromisno Resenje). Compromised weighting method poised of AHP (analytic hierarchy process) and entropy methods

Selection of Elastomer for Robotic Gripper

were used for performing the criteria weighting. These methods were also used for ranking the materials, and the outputs of every method were compared among each other. It was definite that we can implement MCDM methods for the solution of real-time material selection problems. Tungsten carbide–cobalt and Fe–5Cr–Mo–V aircraft steel were found as the best materials for the tool holder production. Cicek et al. [38] proposed to find an accurate solution to the problem of material selection among various materials having numerous properties. This paper proposed a model selection interface to facilitate analytical solutions to various problem concepts in material selection in MADM environments. In particular, the generic framework of the fuzzy axiomatic design-model selection interface (FAD-MSI) was customized and effectively applied to the various material selection problem concepts. Accordingly, the resulting problem–model sets could be referred for the accomplishment of further proposals in the upcoming period.

Figure 6.1 illustrates the inevitable and governing elements of IPMC-assisted micromanipulation strategies. Herein, IPMC being one of the mostly used EAPs has been considered, which can act either individually or in conjugation with a

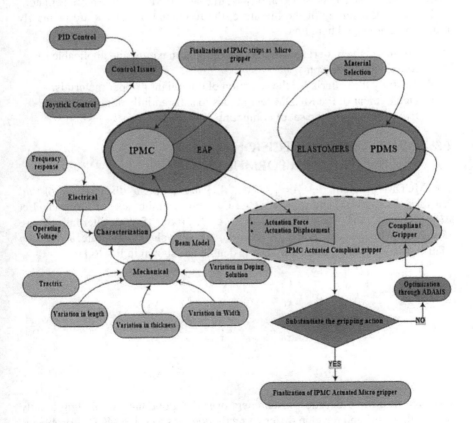

FIGURE 6.1 Issues and aspects of IPMC-assisted micromanipulation.

PDMS-based compliant gripper. Eventually, the reason for selecting PDMS as a compliant robotic material comes into the forefront, which one must substantiate with methodological approach. Characterization of IPMC is another prima facie concern which corroborates both mechanical and electrical aspects. Mechanical aspect includes tractrix (locating the tip position with certainty), equivalent beam model, variation in physical attributes (length, breadth and width) and obviously anticipating the behavior in doping solution. The electrical counterpart of characterization includes the response of IPMC when it is subjected to different frequency and voltage levels. The control issue entails the mathematical modeling so as to fetch the upshots by tweaking the different gains pertinent to PID control. In addition, an attempt has already been reported for joystick control keeping in view of tele-operation.

So far discussed, IPMC actuation is unequivocally meant for primarily aimed at to fetch the following elements (a) actuating force and (b) actuating displacement from operational point of view. Nevertheless, if IPMC is harnessed with PDMS microgripper, compatibility should be adjudged so that the actuating force of IPMC should be sufficient enough to actuate the PDMS gripper. In this regard, optimization of the shape and size may be accomplished to enable IPMC to serve as an actuator.

Based on the survey of the literature, the following objectives are primarily found to be adjudged here unto.

i. To survey the materials which are compliant in nature and applicable for robotic manipulation.
ii. To study the criteria for the selection of compliant gripper materials.
iii. Deployment of the MCDM techniques to substantially adjudge the relative suitability of the set of competing materials.

6.2 MULTI-CRITERIA DECISION MAKING (MCDM) PROBLEM FORMULATION

For MCDM problem, let $A = \{A_1, A_2 \ldots A_m\} (m \geq 2)$ be a discrete set of 'm' feasible alternatives and $C = \{C_1, C_2 \ldots C_n\}$ be a finite set of criteria. Let $M = \{1, 2, \ldots i, \ldots, m\}, N = \{1, 2, \ldots j, \ldots, n\}; i \in M, j \in N$. If '$m$' alternatives are evaluated with respect to the 'n' criteria, whose values constitute a decision matrix, the following decision matrix is denoted as shown in Eq. (6.1).

$$\begin{array}{c c c c c c} & C_1 & \cdots & C_2 & \cdots & C_3 \\ A_1 & x_{11} & \cdots & x_{1j} & \cdots & x_{1n} \\ \cdots & \cdots & \cdots & \cdots & \cdots & \cdots \\ A_2 & x_{i1} & \cdots & x_{ij} & \cdots & x_{in} \\ \cdots & \cdots & \cdots & \cdots & \cdots & \cdots \\ A_3 & x_{m1} & \cdots & x_{mj} & \cdots & x_{mn} \end{array} \quad (6.1)$$

Over the past three decades, researchers over the globe have tried significantly to reveal, unfurl and develop different methodologies to solve MCDM problems. For the present investigation, the authors have made an attempt to develop an

Selection of Elastomer for Robotic Gripper

expert system, obviously encompassing the existing methodologies by harnessing them in a nutshell to mitigate the computational time and related ambiguities. The MCDM methodologies that have been adopted to develop the expert system are shown here under one after one. However, many existing MCDM methodologies are yet to be considered for making the expert system a robust one.

6.3 TECHNIQUE FOR ORDER OF PREFERENCE BY SIMILARITY TO IDEAL SOLUTION

The TOPSIS method was exhibited in Chen and Hwang [39], in concurrence with reference to Hwang and Yoon [2]. According to TOPSIS, the best chosen alternative reveals the closest proximity to the positive-ideal solution and the far-off from the negative-ideal solution according to Yue [1].

The TOPSIS procedure consists of the following steps:

i. Generally, there are benefit attributes and cost attributes in the MCDM problems. In order to measure all attributes in dimensionless units and facilitate inter-attribute comparisons, the normalization is performed and the normalized values r_{ij} are calculated as:

$$r_{ij} = x_{ij} \Big/ \sqrt{\sum_{i=1}^{m} x_{ij}^2} \qquad (6.2)$$

where $j \in B$; B stands for benefit criteria.

$$r_{ij} = 1 - x_{ij} \Big/ \sqrt{\sum_{i=1}^{m} x_{ij}^2} \qquad (6.3)$$

where $j \in C$; C stands for cost criteria.

ii. The weight matrix of criteria is calculated as

$$w = [w_1 \cdots w_j \cdots w_n]; \sum_{j=1}^{n} w_j = 1 \qquad (6.4)$$

iii. The weighted normalized decision matrix is

$$\begin{array}{c c c c c c c} & C_1 & \cdots & C_j & \cdots & C_n \\ A_1 & v_{11} & \cdots & v_{1j} & \cdots & v_{1n} \\ \cdots & \cdots & \cdots & \cdots & \cdots & \cdots \\ A_i & v_{i1} & \cdots & v_{ij} & \cdots & v_{in} \\ \cdots & \cdots & \cdots & \cdots & \cdots & \cdots \\ A_m & v_{m1} & \cdots & v_{mj} & \cdots & v_{mn} \end{array} \qquad (6.5)$$

where $v_{ij} = w_j \times r_{ij}$; v_{ij} refers to the weighted normalized value for the ith alternative subjected to jth criterion.

iv. Determination of positive-ideal solution and negative-ideal solution
The PIS A^+ and NIS A^- are determined as follows:

$$A^+ = \{v_1^+, v_2^+, \ldots, v_n^+\} \text{ and,} \tag{6.6}$$

$$A^- = \{v_1^-, v_2^-, \ldots, v_n^-\} \tag{6.7}$$

where

$$v_j^+ = \max_{1 \leq i \leq m} \{v_{ij}\} (j \in n) \text{ and } v_j^- = \min_{1 \leq i \leq m} \{v_{ij}\} (j \in n).$$

v. Measure the distance from positive- and negative-ideal solutions.
The dispersion of each alternative from PIS, S_i^+, is given as

$$S_i^+ = \left(\sum_{j=1}^{n} \left(v_j^+ - v_{ij} \right)^2 \right)^{1/2}, i \in m \tag{6.8}$$

Similarly, the dispersion of each alternative from NIS, S_i^-, is given as

$$S_i^- = \left(\sum_{j=1}^{n} \left(v_{ij} - v_j^- \right)^2 \right)^{1/2}, i \in m \tag{6.9}$$

vi. Calculation of the closeness coefficients to the ideal solutions.
The **closeness coefficient** of the ith alternative A_i with respect to ideal solution is defined as

$$C_i = S_i^+ / \left(S_i^+ + S_i^- \right), i \in m \tag{6.10}$$

Since $S_i^+ \geq 0 \, (i \in m)$ and $S_i^- \geq 0 \, (i \in m)$, then, clearly, $C_i \in [0,1]$, $i \in m$.

vii. Ranking of the preference order.
Alternatives poised with larger C_i mean that they are juxtaposed to the PIS and hence they can be prioritized according to the descending order of C_i.

6.4 COMPLEX PROPORTIONAL ASSESSMENT

COPRAS was developed by Zavadskas et al. [40], according to which the project's complex efficiency directly depends on the values and weights considered in the project Antucheviciene et al. [12]. Among the ideal and ideal-worst solutions, this method selects the best decision Chatterjee et al. [14]. The COPRAS method is generally used for evaluating the solution and selecting the best alternative among a set of available alternatives and criteria for a particular engineering problem.

Selection of Elastomer for Robotic Gripper

The COPRAS method consists of the following steps.

i. Normalize the decision matrix by the following equation which will give the dimensionless values.

$$R = \left[r_{ij} \right]_{m \times n} = \frac{x_{ij}}{\sum_{i=1}^{m} x_{ij}} \quad (6.11)$$

ii. Calculate the weighted normalized decision matrix, D.

$$D = \left[y_{ij} \right]_{m \times n} = r_{ij} \times w_j \quad (6.12)$$

where r_{ij} is the normalized performance value and w_j is the weight of the respective criteria such that

$$\sum_{i=1}^{m} y_{ij} = w_j \quad (6.13)$$

The normalization is owing to a particular fashion so that w_j is the weight of the jth criterion which is proportionally distributed among all the alternatives under jth criterion as shown in Eq. (6.13).

iii. For both benefit and cost criteria, the sums of weighted normalized values are calculated using the following equations:

$$S_{+i} = \sum_{j=1}^{k} y_{+ij} \quad (6.14)$$

$$S_{-i} = \sum_{j=1}^{n-k} y_{-ij} \quad (6.15)$$

where y_{+ij} and y_{-ij} are the weighted normalized values for the benefit and cost criteria, respectively, and k is the number of benefit criteria. The best alternative is the one with the higher value of S_{+i} and the lower value of S_{-i}. S_{+i} and S_{-i} determine the index of attainment toward the goal by each alternative.

The sum of S_{+i} and S_{-i} for all the alternatives is equal to the sum of weights of all the benefit and cost criteria as given by Eqs. (6.14) and (6.15).

$$S_{+i} = \sum_{i=1}^{m} S_{+i} = \sum_{i=1}^{m} \sum_{j=1}^{k} y_{+ij} \quad (6.16)$$

$$S_{-i} = \sum_{i=1}^{m} S_{-i} = \sum_{i=1}^{m} \sum_{j=1}^{n-k} y_{-ij} \quad (6.17)$$

iv. Relative significances of the alternatives are calculated on the basis of Q_i. Significance or priority of an alternative depends upon Q_i. So the higher the value of Q_i, the better the alternative. So, the alternative with the highest value of Q_{max} is the best choice among the alternatives.

The relative significance value Q_i for an alternative is calculated by the equation given below:

$$Q_i = S_{+i} + \frac{S_{-\min} \sum_{i=1}^{m} S_{-i}}{S_{-i} \sum_{i=1}^{m} \left(S_{-\min} / S_{-i} \right)} \quad (i = 1, 2, \ldots, m) \tag{6.18}$$

where $S_{-\min}$ is the minimum value of S_{-i}.

v. The degree of alternative's utility (U_i) is used to calculate complete ranking of the alternative, which depends upon the relative significance value (Q_i). So, the quantity utility (U_i) is calculated using the equation given below:

$$U_i = \left[\frac{Q_i}{Q_{\max}} \right] \times 100\% \tag{6.19}$$

where $Q_{\max}$ is the highest relative significance value. The degree of utility for the alternatives varies from 0% to 100%. Thus, **COPRAS** is used to calculate degree of utility of different alternatives by which we can calculate the best alternative in a decision making problem involving multiple criteria and alternatives.

6.5 MULTI-OBJECTIVE OPTIMIZATION ON THE BASIS OF RATIO ANALYSIS

The MOORA method was first introduced by Brauers [41] and later on advocated by Brauers and Zavadskas [8] as a MCDM tool to find the best alternative among a set of alternatives available on the basis of several criteria. According to this method, the performance of an alternative is evaluated by measuring the difference between the weighted sum of benefit contribution and its cost contribution. This dispersion eventually gives rise to composite score, which is the measure of suitability of a particular alternative to its competing counterparts.

The different steps in the MOORA method are given below:

i. To convert the prima facie decision matrix into rational dimensionless stature, Eq. (6.20) is employed to obtain normalized decision matrix.

$$r_{ij} = \frac{x_{ij}}{\sqrt{\sum_{i=1}^{m} x_{ij}^2}} \tag{6.20}$$

where r_{ij} is the normalized version corresponding to x_{ij} and eventually unfurls a range $0 < r_{ij} < 1$.

ii. Now to find the final ranking of the alternatives for a MCDM problem, the weights of the different criteria w_j are calculated and multiplied with the normalized values, i.e., r_{ij}.

$$v_{ij} = w_j \times r_{ij} \tag{6.21}$$

Selection of Elastomer for Robotic Gripper

iii. For each alternative, the total composite score is calculated using Eq. (6.22) and expressed as

$$y_i = \sum_{j=1}^{k} v_{ij} - \sum_{j=k+1}^{n} v_{ij} \qquad (6.22)$$

where k is the number of benefit criteria. Here, y_i may owe to positive or negative values depending on the number of cost and benefit criteria in the problem.

iv. Ranking based on y_i is the final preference. So, the alternative with the highest value y_{max} is the best alternative.

6.6 ELECTRE II

ELECTRE II developed by Benayoun et al. [42] is extensively used for making decision by decision makers across the globe. It is among one of the most used outranking methods. Shanian et al. [43] used this method for suitable material selection in the loaded thermal conductor.

Combined with AHP, a decision model is proposed by Zhang and Chen [13] for the selection of optimal design scheme for the computer numerical control (CNC) machine.

The procedural steps for this method are mentioned below:

i. Normalize the decision matrix to get the dimension-independent data using the following equation:

$$r_{ij} = \frac{x_{ij}}{\sqrt{\sum_{j=1}^{n} x_{ij}^2}} \quad (i \in m, j \in n) \qquad (6.23)$$

ii. The weight matrix of criteria is calculated as

$$w =; \lfloor w_1 \ldots w_j \ldots w_n \rfloor; \sum_{j=1}^{n} w_j = 1 \qquad (6.24)$$

iii. Construct the weighted normalized decision matrix using the following:

$$v_{ij} = w_i \times r_{ij} \, (i \in m, j \in n) \qquad (6.25)$$

Concordance matrix

$$C_{ii'} = \sum_{j=1}^{n} w_j \qquad (6.26)$$

Here 'j' is a set of criteria in which 'i' is preferred over 'i'.

The concordance matrix is

$$C = \begin{bmatrix} - & c_{12} & \cdots & c_{1m} \\ c_{21} & - & \cdots & c_{2m} \\ \cdots & \cdots & \cdots & \cdots \\ c_{m1} & c_{m2} & \cdots & - \end{bmatrix} \quad (6.27)$$

The discordance matrix is

$$D(a,b) = \max_j \left(g_j(b) - g_j(a) \right) / \delta \quad (6.28)$$

where $g_j(b) - g_j(a)$ is the difference in the performance of alternatives a and b for criterion j. Here, j belongs to the criterion in which alternative b is preferred over alternative a.

δ is the maximum absolute difference between a and b considering all criteria.

The discordance interval matrix

$$D = \begin{bmatrix} - & d_{12} & \cdots & d_{1m} \\ d_{21} & - & \cdots & d_{2m} \\ \cdots & \cdots & \cdots & \cdots \\ d_{m1} & d_{m2} & \cdots & - \end{bmatrix} \quad (6.29)$$

iv. Computation of concordance interval matrix

$$\bar{c} = \sum_{a=1}^{m} \sum_{b}^{m} \frac{c(a,b)}{m(m-1)} \quad (6.30)$$

The Boolean matrix E is

$$\begin{cases} e(a,b) = 1 & \text{if} \quad c(a,b) \geq \bar{c} \\ e(a,b) = 0 & \text{if} \quad c(a,b) < \bar{c} \end{cases} \quad (6.31)$$

where $\bar{c}$ is the critical value determined by the concordance interval matrix.

v. Computation of discordance index which is the presence of dissatisfaction.

$$\bar{d} = \sum_{a=m}^{m} \sum_{b}^{m} \frac{d(a,b)}{m(m-1)} \quad (6.32)$$

Discordance index matrix

$$\begin{cases} f(a,b) = 1 & \text{if} \quad d(a,b) \geq \bar{d} \\ f(a,b) = 0 & \text{if} \quad d(a,b) < \bar{d} \end{cases} \quad (6.33)$$

Selection of Elastomer for Robotic Gripper 135

vi. c_a and d_a are the net superior and inferior values. c_a be sums together the number of competitive superiority for all alternatives

$$c_a = \sum_{b=1}^{n} c_{(a,b)} - \sum_{b=1}^{n} c_{(b,a)} \qquad (6.34)$$

d_a is used to determine the number of inferiority ranking the alternatives

$$d_a = \sum_{b=1}^{n} d_{(a,b)} - \sum_{b=1}^{n} d_{(b,a)} \qquad (6.35)$$

6.7 DETERMINATION OF ENTROPY WEIGHT

Entropy ascertains the randomness in terms of uncertainty of the information employing probability theory. It indicates that a data set having wide range or distribution is having more uncertainty than a sharply peaked one Rao [11]. Equation (6.36) shows the decision matrix 'A' with 'm' alternatives and 'n' criteria (Chou et al. [44]):

$$A = \begin{bmatrix} x_{11} & x_{12} & \cdots & x_{1n} \\ x_{21} & x_{22} & \cdots & x_{2n} \\ \cdots & \cdots & \cdots & \cdots \\ x_{m1} & x_{m2} & \cdots & x_{mn} \end{bmatrix} \qquad (6.36)$$

where x_{ij} ($i = 1, 2,\ldots, m; j = 1, 2, \ldots, n$) is the performance value of the ith alternative to the jth criterion.

In order to obtain the criteria weights by the entropy method, the data set is normalized employing Eq. (6.37) [11]. The prima facie concern of normalization is to make the data set dimensionless, and through normalization, the data set is compelled to confine within the range of (0,1).

$$p_{ij} = \frac{x_{ij}}{\sqrt{\sum_{i=1}^{m} x_{ij}^2}} \qquad (6.37)$$

Afterward, the entropy value E_j of jth criterion can be obtained as:

$$E_{j'} = k \sum_{i=1}^{n} p_{ij} \ln(p_{ij}); j \in n \qquad (6.38)$$

where $k = -1/\ln(m)$, a constant that guarantees $0 \leq E_j \leq 1$. The degree of divergence (d_j) of the normalized sequence of each criterion can be determined from Eq. (6.39).

$$d_j = |1 - E_j| \qquad (6.39)$$

Thus, the weight of entropy of jth criterion can be defined as:

$$w_j = \frac{d_j}{\sqrt{\sum_{i=1}^{m} d_j}} \qquad (6.40)$$

6.8 SPEARMAN'S RANK CORRELATION COEFFICIENT

The Spearman's rank correlation coefficient [http://www.bws.wilts.sch.uk/] 'r_s' is a reliable and fairly simple method of testing both the strength and direction (positive or negative) of any correlation between two variables.

$$\begin{array}{cccc} \text{Alternatives} & x_i & y_i & d_i^2 \\ A_1 & x_1 & y_1 & (x_1 - y_1)^2 \\ A_2 & x_2 & y_2 & (x_2 - y_2)^2 \\ A_i & x_i & y_i & (x_i - y_i)^2 \\ A_m & x_m & y_m & (x_m - y_m)^2 \end{array} \tag{6.41}$$

where x_i and y_i are the ranks of different alternatives through two different MCDM methods. d_i is the difference between the ranks achieved by two MCDM techniques for ith alternative, where ($i \in m$).

Hence, Spearman's rank correlation coefficient (r_s) is defined by the following equation:

$$r_s = 1 - \left(\frac{6 \sum d^2}{m^3 - m} \right) \tag{6.42}$$

The value of 'r_s' should be between −1 (perfect negative correlation) and +1 (perfect positive correlation). The nearer the value is to 0, the weaker the correlation.

6.9 CRITERIA FOR COMPLIANT MATERIAL SELECTION

The selection of optimal material for making a compliant robot gripper among various materials having many attributes is the basis of any MCDM problem. The best material is the one which serves the intended purpose for a desired period of time under the given condition. The easy availability of the material in the market and in sufficient quantity to meet the requirement is also an important issue. Hence, the best material having appropriate balance between the mechanical properties such as hardness, density, elongation at break, tensile stress and cost of the material is to be chosen. The factors that should be taken into consideration for making the compliant gripper are as follows.

6.9.1 Hardness

Hardness is the resistance to permanent deformation. Hardness is a measure of the force (or stress) that needs to be applied to a material if it is desired that some deformation remains when the force is taken away. It is generally used in the context of indentation, so if the material is not hard enough, it would get a scratch or permanent indent. As both hardness and stiffness come from inter-atomic forces in the material, high hardness tends to be correlated with high stiffness. As high stiffness is necessary while designing the compliant robotic gripper such that localized deformation doesn't occur, a high hardness value is solicited for the present case.

6.9.2 Density

For a gripper to grip firmly, a necessary attribute that the authors have taken up while solving the MCDM problem is density. The density of polymer materials used for making the gripper must have a nominal value. That is, neither it must be a high value, nor should it be less.

High values of density will mean higher intermolecular forces of attraction and thus high rigidity of the compliant robotic gripper. Therefore, the gripper shall become as stiff as the force will not be transmitted and thus the basic intended purpose of gripping shall not be sufficed. A lesser density of the gripper will mean the basic strength of the gripper shall be so less that it will not be able to hold any object. The nominal value is hence decided such that the gripper might be able to have a sufficient load-bearing capacity and shall not give away while grasping any object.

6.9.3 Tensile Strength

The tensile strength of a material is essentially its ability to withstand tensile loads without failure. For the present investigation, the compliant gripper design includes two numbers of hidden four-bar mechanism to activate the jaws. Eventually, each of the four-bar linkages has two ground points which are fixed and about which the compliant gripper flexes when it is in action. Now if the tensile strength is more, then the resistance will also be more; therefore, the IPMC strips won't be able to significantly flex the compliant gripper to ensure gripping action. On the other hand, if the tensile strength is low, then the flexion will be accomplished at ease, but if the deformation is permanent, then the input port will never be able to come back to its initial position. Hence, the nominal value of tensile strength is taken for material selection problem for compliant gripper.

6.9.4 Elongation at Break

Elongation at break signifies the degree of deformation during rupture, which is also known as fracture strain. A higher value of it means the material is non-elastic, and vice versa. Compliant mechanisms work on the notion of force propagation rather than localized deformation. Therefore, if the elongation at break is more, it will imply a higher rate of localized deformation at the cost of force propagation. Owing to these reasons, a lower value of elongation at break is chosen.

6.9.5 Cost

For every application, there is a cost limit which a designer cannot breach. Whenever that limit is reached, we have to look for another alternative. Hence, we have taken cost criterion as lower the best. The cost of the material should be low so that the overall processing operation is economical to the manufacturer. Apart from being economical, the material's availability in the market is also an important concern.

The material should be sufficiently available in the market to meet the requirement of the manufacturer.

6.10 COMPLIANT MATERIALS

6.10.1 Ethylene–Propylene Diene Monomer (EPDM)

As per classification of ASTM, EPDM belongs to M class elastomer, which has a standard chain of the polyethylene. The percentage of ethylene in EPDM is approximately 45%–50%. For obtaining comparatively good mixing and extrusion, a high percentage of ethylene is required as it improves the loading capability of the polymer. The dienes comprise 2.5%–12% by weight of the entire EPDM composition. The basic function of diene is to provide resistance to unwanted tackiness and creep. It also shows an exceptional electrical insulation.

6.10.2 Ethylene–Vinyl Acetate (EVA)

EVA consists of ethylene and vinyl acetate, in which the content of vinyl acetate is limited up to 40% starting from 10% by weight, wherein the remainder being ethylene. EVA belongs to elastomeric polymer category which looks like rubber, and it is very soft and flexible by virtue of its constitution. EVA extends good low-temperature toughness, good transparency and aesthetics, good crack propagation resistance and excellent hot-melt bonding waterproof properties. EVA also has a characteristic vinegar-like smell and is highly recommended for many electrical applications.

6.10.3 Ethylene–Propylene Monomer (EPM)

Ethylene–propylene-based rubbers entirely consist of ethylene and propylene monomers which are the primal form of any non-polar synthetic rubber. EPM rubbers do not possess carbon–carbon double bondage. The performance of EPDM materials can be highly improved with the help of sulfur or peroxide. The addition of sulfur highly helps in the improvement of the mechanical properties, and peroxide upgrades the heat stability of the EPM materials. The polymer chains of EPM consist of entirely saturated hydrocarbon backbones, owing to which it exhibits exceptional resistance to heat and oxidation.

Selection of Elastomer for Robotic Gripper

6.10.4 Polydimethylsiloxane (PDMS)

Polydimethylsiloxane (PDMS) is a polymeric organ silicon compound, which is also referred to as silicone. PDMS is an abundantly used silicon-based organic polymer and is better known for its flow properties. PDMS is transparent in nature and is generally inert, non-toxic and inflammable. PDMS is viscoelastic, which means it acts as a sticky fluid for a long flow times and at high temperatures. PDMS rubber also acts as an elastic solid for short flow time.

6.10.5 Polyurethane (PU)

Polyurethane (PUR and PU) is a polymer comprising of organic units that are joined together by carbamate (urethane) links. Polyurethane polymers are generally formed by the reaction between a di- or polyisocyanate and a polyol. Polyurethane products are generally referred to as 'urethanes'. Polyurethanes are extensively used for the manufacture of high-resilience foam seating, durable elastomeric wheels and tires, electrical potting compounds and high-performance adhesives.

6.10.6 Ethylene–Propylene Terpolymer (EPT)

EPT is a synthetic rubber primarily known for its exceptional capability to adjust according to various climatic conditions and its high resistance to ozone, heat and coldness. It also performs well as an electric insulator and is known for its chemical resistance. It has widely been used in automotive parts, electric wires and cables, as well as many other industrial products owing to its good heat-resistant properties. It also exhibits good electric insulation and other comprehensive properties.

6.10.7 Polyvinylidene Fluoride (PDVF)

Polyvinylidene fluoride (PDVF) is produced by the polymerization of vinylidene difluoride, which is highly non-reactive and purely a thermoplastic fluoropolymer. PDVF is a polymer that is used generally in applications demanding higher strength and resistance to solvents, acids, bases and heat.

6.11 RESULTS

To substantially adjudge the suitability of the best compliant material for the designing of soft robotic gripper among a set of compliant materials, the authors undertook several MCDM techniques such as TOPSIS, COPRAS, MOORA and ELECTRE II to calculate the same. Seven different compliant materials were taken into consideration to solve the material selection problem based on five selection criteria, namely hardness, density, tensile strength, percentage elongation at break and cost. The decision matrix for the material selection problem is shown in Table 6.1.

TABLE 6.1
Initial Decision Matrix

Material	Hardness (Shore)	Tensile Strength (MPa)	Elongation at Break (%)	Density (g/cc)	Cost ($/kg)
EPDM	65	13.5	400	1.01	1.585
EPM	62.5	14	350	0.86	1.629
EVA	80	19	550	0.945	1.275
PDMS	70	13	65	0.965	0.25
PT	62.5	38.5	575	1.25	1.125
EPT	60	12.5	545	0.85	2.69
PDVF	64.5	43	200	1.75	0.2025

TABLE 6.2
Normalized Decision Matrix

Material	Hardness	Tensile Strength	Elongation at Break	Density	Cost
EPDM	0.3686	0.2036	0.3568	0.3380	0.4039
EPM	0.3544	0.2112	0.3122	0.2878	0.4152
EVA	0.4537	0.2866	0.4906	0.3162	0.3249
PDMS	0.3969	0.1961	0.0580	0.3229	0.0637
PT	0.3544	0.5807	0.5129	0.4183	0.2867
EPT	0.3402	0.1885	0.4861	0.2844	0.6855
PDVF	0.3658	0.6486	0.1784	0.5856	0.0516

TABLE 6.3
Entropy Value, Degree of Divergence and Criteria Weights

	Hardness	Tensile Strength	Elongation at Break	Density	Cost
E_j	1.3172	1.1517	1.1543	1.2795	1.0493
d_j	0.3172	0.1517	0.1543	0.2795	0.0493
w_j	0.3332	0.1593	0.1621	0.2936	0.0518

6.11.1 Weight Determination by Entropy Method

Step 1: The initial decision matrix is normalized as per Eq. (6.37), and the resulting matrix is shown in Table 6.2.

Step 2: The entropy value is determined as per Eq. (6.38), the degree of divergence is calculated as per Eq. (6.39), and finally, the entropy weight is calculated as per Eq. (6.40) for each attribute. The formulated table is shown in Table 6.3.

Selection of Elastomer for Robotic Gripper

6.11.2 SOLUTION BY TOPSIS

Step 1: The normalized matrix is calculated, and the table obtained is shown in Table 6.2.

Step 2: After calculating the normalized matrix as above, the entropy method is used to determine entropy weights for each attribute and the formulated table is shown in Table 6.4.

Step 3: The Euclidean distances from positive-ideal solution (PIS) and negative-ideal solution (NIS) are ascertained, and finally, the rank order for the elements is shown in Table 6.5.

6.11.3 SOLUTION BY COPRAS

Step 1: The decision matrix is normalized as per Eq. (6.11) to obtain dimensionless values of different criteria so that all the alternatives can be compared as shown in Table 6.2.

Step 2: With the help of the entropy method, the weights are ascertained as per Eqs. (6.12) and (6.13), which are proportionally distributed among

TABLE 6.4
Weighted Normalized Matrix

Material	Hardness	Tensile Strength	Elongation at Break	Density	Cost
EPDM	0.0466	0.1567	0.0242	0.2575	0.0094
EPM	0.0448	0.1541	0.0211	0.2804	0.0096
EVA	0.0574	0.1280	0.0332	0.2789	0.0075
PDMS	0.0502	0.1593	0.0039	0.2723	0.0015
PT	0.0448	0.0261	0.0347	0.1791	0.0067
EPT	0.0430	0.1567	0.0329	0.2759	0.0159
PDVF	0.0463	0.0026	0.0121	0.0147	0.0012

TABLE 6.5
Euclidean Distances and Closeness Coefficients

	PIS S^+	NIS S^-	Closeness Coefficient	Rank
EPDM	0.0335	0.2879	0.8957	3
EPM	0.0235	0.3062	0.9287	2
EVA	0.0434	0.2929	0.8710	5
PDMS	0.0108	0.3036	0.9658	1
PT	0.1707	0.1663	0.4935	6
EPT	0.0359	0.3033	0.8942	4
PDVF	0.3088	0.0272	0.0809	7

S^+ is the distance from PIS, and S^- is the distance from NIS.

all the alternatives, and the weighted normalized matrix is calculated, which is shown in Table 6.4.

Step 3: The sums of weighted normalized values are calculated for beneficial and non-beneficial attributes. Further, the significances of the alternatives are determined on the basis of defining the positive attributes as S^+ and negative attributes as S^- as per Eqs. (6.16) and (6.17). Finally, the relative significance value (Q) is ascertained as shown in Table 6.6, which thereupon helps us in deciding the rank order for the alternatives.

6.11.4 Solution by MOORA

Step 1: The normalized matrix is calculated to convert the decision matrix into non-biased matrix, and the resulting table is shown in Table 6.2.

Step 2: The weights of different alternatives are calculated and multiplied with the normalized value to determine the weighted normalized matrix as shown in Table 6.4.

Step 3: The total composite score is calculated as per Eq. (6.22), and finally, the rank for the alternatives is decided as shown in Table 6.7.

TABLE 6.6
Relative Significance Values for Ranking of the Materials in COPRAS

S^+	S^-	Relative Significance Q	Rank
0.4608	0.0335	0.4771	5
0.4793	0.0308	0.4970	2
0.4643	0.0407	0.4776	4
0.4819	0.0054	0.5829	1
0.2500	0.0414	0.2632	6
0.4757	0.0488	0.4869	3
0.0635	0.0133	0.1047	7

TABLE 6.7
Composite Score by MOORA

$\sum_{j=1}^{k} v_{ij}$	$\sum_{j=k+1}^{n} v_{ij}$	Composite Score y_i	Rank
0.460808	0.033521	0.427287	3
0.479299	0.030762	0.448537	2
0.464254	0.040746	0.423508	5
0.481892	0.005403	0.476489	1
0.250024	0.041369	0.208656	6
0.475716	0.048808	0.426907	4
0.063546	0.013273	0.050273	7

Selection of Elastomer for Robotic Gripper

6.11.5 Solution by ELECTRE II

Step 1: The normalized matrix is calculated such that all the attributes are made higher the best, and the table obtained is shown in Table 6.2.

Step 2: The weights are determined as per the entropy method, which are proportionally distributed among all the alternatives, and the weighted normalized matrix is calculated, which is shown in Table 6.4.

Step 3: The concordance interval matrix is calculated as per Eq. (6.30), and the net superior value c_a is obtained as per Eq. (6.34), which helps us in determining the rank order of the alternatives as per Table 6.8.

Step 4: The discordance interval matrix is calculated as per Eq. (6.32), and the net superior value d_a is obtained as per Eq. (6.35), which helps us in determining the rank order of the alternatives as per Table 6.9.

TABLE 6.8
Concordance Matrix for Concordance Ranking by ELECTRE II

	EPDM	EPM	EVA	PDMS	PT	EPT	PDVF	Sum	c_a	Rank
EPDM		0.544	0.321	0	0.789	0.6265	0.492	2.7725	−0.455	4
EPM	0.456		0.615	0.294	0.7815	0.841	0.453	3.4405	0.881	3
EVA	0.679	0.385		0.627	0.948	0.627	0.786	4.052	2.104	2
PDMS	1	0.706	0.373		1	0.706	0.948	4.733	3.466	1
PT	0.211	0.2185	0.052	0		0.385	0.453	1.3195	−3.361	7
EPT	0.3735	0.159	0.373	0.294	0.615		0.453	2.2675	−1.465	6
PDVF	0.508	0.547	0.214	0.052	0.547	0.547		2.415	−1.17	5
Sum	3.2275	2.5595	1.948	1.267	4.6805	3.7325	3.585			

TABLE 6.9
Discordance Matrix for Discordance Ranking by ELECTRE II

	EPDM	EPM	EVA	PDMS	PT	EPT	PDVF	Sum	d_a	Rank
EPDM		0.262	0.237	−0.012	0.568	0.416	0.515	1.984	−0.253	4
EPM	0.574		0.230	0.038	0.525	0.364	0.537	2.269	0.550	3
EVA	0.515	0.643		0.163	0.369	0.705	0.496	2.891	1.536	2
PDMS	0.649	0.622	0.558		0.331	0.621	0.461	3.243	3.088	1
PT	0.012	0.012	0.004	−0.072		0.042	0.600	0.598	−1.884	6
EPT	0.390	0.088	0.243	0.021	0.532		0.507	1.782	−0.458	5
PDVF	0.096	0.092	0.082	0.018	0.156	0.093		0.537	−2.578	7
Sum	2.237	1.719	1.355	0.156	2.482	2.241	3.116			

TABLE 6.10
Aggregate Rank by ELECTRE II

Rank Obtained by Concordance	Rank Obtained by Discordance	Aggregate Rank
4	4	4
3	3	3
2	2	2
1	1	1
6	7	7
5	6	5
7	5	6

Step 5: The aggregate rank is calculated by averaging the rank attained by concordance as well as discordance interval matrices, and the final rank is shown in Table 6.10.

6.12 DISCUSSION

Owing to the prima facie concern of the present problem statement, an attempt was made to assimilate the rank orders laid by different MCDM methodologies. As per TOPSIS methodology the rank order fetched by the different compliant materials is: PDMS>EPM>EPDM>EPT>EVA>PT>PDVF. As per COPRAS, the rank order fetched by the different compliant materials is: PDMS>EPM>EPT>EVA>EPDM>EPT>PDVF. As per MOORA, the rank order fetched by the different compliant materials is: PDMS>EPM>EPDM>EPT>EVA>PT>PDVF. As per ELECTRE II, the rank order fetched by the different compliant materials is: PDMS>EVA>EPM>EPDM>EPT>PDVF>PT. It is evident from Figure 6.2 that PDMS always fetches the top rank in all the methodologies, whereas PDVF fetches the least rank in all the methods except for ELECTRE II, wherein we observe that PT obtains the least rank. It may be clearly noted that, although there is a trivial change in the least rank, the corresponding rank orders remain the same and do not stagger much, thus attaining stability.

Table 6.11 depicts the rank correlation matrix in between the different MCDM methodologies. It has been observed that the maximum rank correlation exists between COPRAS and TOPSIS as well as MOORA and COPRAS.

6.13 CONCLUSIONS

In the present investigation, the authors took up the problem of selecting the best material for the design of a compliant robotic gripper based on certain MCDM techniques. Seven different compliant materials were taken, which were ethylene–propylene diene monomer (EPDM), ethylene–vinyl acetate (EVA), ethylene–propylene monomer (EPM), polydimethylsiloxane (PDMS), polyurethane (PU)

Selection of Elastomer for Robotic Gripper

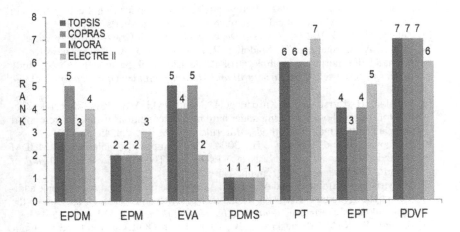

FIGURE 6.2 Comparison of rank order by different methods.

TABLE 6.11
Rank Correlation Matrix

TOPSIS	COPRAS	MOORA	ELECTRE II	
–	89.29	100	75	TOPSIS
	–	89.29	78.57	COPRAS
		–	75	MOORA
			–	ELECTRE II

and polyvinylidene fluoride (PDVF). Applications of these materials in compliant gripper design were found during the literature review. Five different criteria were taken up, which were hardness, tensile strength, density, elongation at break and cost, to ascertain the relative suitability of the materials. Certain MCDM techniques were employed, such as TOPSIS, COPRAS, MOORA and ELECTRE II, to obtain the material which is best suitable to simultaneously attain the criteria mentioned earlier. PDMS was found as the best material from application point of view as revealed by all the methods.

In the next chapter, the reader will find a new research area and methodology to work with polar region atmospheric electric field impact on human beings using ionic polymer–metal composites (IPMCs).

REFERENCES

[1] Yue, Z., 2011. A method for group decision-making based on determining weights of decision makers using TOPSIS, *Applied Mathematical Modelling*, Vol. 35, pp. 1926–1936.

[2] Hwang, C. L., Yoon, K., 1981. Multiple attribute decision making. Lecture Notes in Economics and Mathematical Systems, No. 186. Springer-Verlag, Berlin.
[3] Zeleny, M., 1982. *Multiple Criteria Decision Making*, McGraw-Hill, New York.
[4] Milani, A. S., Shanian, A., Madoliat, R. and Nemes, J. A., 2005. The effect of normalization norms in multiple attribute decision making models: a case study in gear material selection, *Structural and Multidisciplinary Optimization*, Vol. 29, pp. 312–318.
[5] Caliskan, H., Kursuncu, B., Kurbanogclu, C. and Sevki, Y.G., 2013. Material selection for the tool holder working under hard milling conditions using different multi criteria decision making methods, *Materials and Design*, Vol. 45, pp. 473–479.
[6] Opricovic, S. and Tzeng, G. H., 2004. The Compromise solution by MCDM methods: a comparative analysis of VIKOR and TOPSIS, *European Journal of Operational Research*, Vol. 156, pp. 445–455.
[7] Memariani, A., Amini, A. and Alinezhad, A., 2009. Sensitivity analysis of simple additive weighting method (SAW): the results of change in the weight of one attribute on the final ranking of alternatives, *Journal of Industrial Engineering*, Vol. 4, pp. 13–18.
[8] Brauers, W. K. M., Zavadskas, E. K., 2006. The MOORA method and its application to privatization in a transition economy. *Control and Cybernetics*, Vol. 35, pp. 445–469.
[9] Karande, P. and Chakraborty, S., 2012. Application of multi-objective optimization on the basis of ratio analysis (MOORA) method for materials selection. *Materials and Design*, Vol. 37, pp. 317–324.
[10] Maniya, K. and Bhatt, M. G., 2010. A selection of material using a novel type decision-making method: preference selection index method, *Materials and Design*, Vol. 31, pp. 1785–1789.
[11] Rao, R. V. and Padmanabhan, K. K., 2007. Rapid prototyping process selection using graph theory and matrix approach, *Journal of Material Processing Technology*, Vol. 194, pp. 81–88.
[12] Antucheviciene, J., Zakarevicius, A., Zavadskas, E. K., 2011. Measuring congruence of ranking result applying particular MCDM methods. *Informatica*, Vol. 23, pp. 319–338.
[13] Pang, J., Zhang, G. and Chen, G., 2011. ELECTRE I decision model of reliability design scheme for computer numerical control machine, *Journal of Software*, Vol. 6, No. 5, pp. 894–900.
[14] Chatterjee, P., Athawale, V. M., Chakraborty, S., 2011. Materials selection using complex proportional assessment and evaluation of mixed data methods, *Materials and Design*, Vol. 32, pp. 851–860.
[15] Chatterjee, P. and Chakraborty, S., 2012. Material selection using preferential ranking methods, *Materials and Design*, Vol. 35, pp. 384–393.
[16] Roth, R., Field, F. and Clark, J., 1993. Materials selection and multi-attribute utility analysis, *Journal of Computer Aided Materials Design*, Vol. 1, pp. 325–342.
[17] Prasad, K. and Chakraborty, S., 2013. A quality function deployment-based model for material selection, *Materials and Design*, Vol.49, pp. 525–535.
[18] Manshadi, B. D., Mahmudi, H., Abedian, A. and Mahmudi, R., 2007. A novel method for materials selection in mechanical design: combination of non-linear normalization and a modified digital logic method, *Materials and Design*, Vol. 28, pp. 8–15.
[19] Athawale, V. M., Kumar, R. and Chakraborty, S., 2011. Decision making for material selection using the UTA method, *International Journal of Advance Manufacturing Technology*, Vol. 57, pp. 11–22.
[20] Parkan, C. and Wu, M. L., 2000. Comparison of three modern multi criteria decision-making tools, *International Journal of Systems Science*, Vol. 31, pp. 497–517.

[21] Jahan, A., Ismail, M. Y., Mustapha, F. and Sapuan, S. M., 2010. Material selection based on ordinal data, *Materials and Design*, Vol. 3, pp. 3180–3187.
[22] Reddy, A. N., Maheshwari, N., Sahu, D. K., and Ananthasuresh, G. K., 2010. Miniature compliant grippers with vision-based force sensing, *IEEE Transactions on Robotics*, Vol. 26, No. 5, pp. 867–877.
[23] Zubir, M. N. M., Shirinzadeh, B. and Tian, Y., 2009. A new design of piezoelectric driven compliant-based microgripper for micromanipulation, *Mechanism and Machine Theory*, Vol. 44, pp. 2248–2264.
[24] Changa, R. J., Wang, H. S. and Wang, Y. L., 2003. Development of mesoscopic polymer gripper system guided by precision design axioms, *Precision Engineering*, Vol. 27, pp. 362–369.
[25] Zhang, D., Zhang, Z., Gao, Q., Xu, D. and Liu, S., 2015. Development of a monolithic compliant SPCA-driven micro-gripper, *Mechatronics*, Vol. 25, pp. 37–43.
[26] Voigt, D., Karguth, A. and Gorb, S., 2012. Shoe soles for the gripping robot: Searching for polymer-based materials maximising friction, *Robotics and Autonomous Systems*, Vol. 60, pp. 1046–1055.
[27] Bhattacharya, S., Chattaraj, R., Das, M., Patra, A., Bepari, B. and Bhaumik, S., 2015. Simultaneous parametric optimization of IPMC actuator for compliant gripper, *International Journal of Precision Engineering and Manufacturing*, Vol. 16, No. 11, pp. 2289–2297.
[28] Bahraminasab, M. and Jahan, A., 2011. Material selection for femoral component of total knee replacement using comprehensive VIKOR, *Materials and Design*, Vol. 32, pp. 4471–4477.
[29] Alemi, M., Jalalifar, H., Kamali, G. R. and Kalbasi M., 2011. A mathematical estimation for artificial lift systems selection based on ELECTRE model, *Journal of Petroleum Science and Engineering*, Vol. 78, pp. 193–200.
[30] Jee, D. H. and Kang, K. J., 2000. A method for optimal material selection aided with decision making theory, *Materials and Design*, Vol. 21, pp. 199–206.
[31] Rao, R. V. and Patel, B. K., 2010. A subjective and objective integrated multiple attribute decision making method for material selection, *Materials and Design*, Vol. 31, pp. 4738–4747.
[32] Jahan, A., Bahraminasab, M. and Edwards, K. L., 2012. A target-based normalization technique for materials selection, *Materials and Design*, Vol. 35, pp. 647–654.
[33] Jahan, A., Ismail, M. Y., Shuib, S., Norfazidah, D. and Edwards, K. L., 2011. An aggregation technique for optimal decision-making in materials selection, *Materials and Design*, Vol. 32, pp. 4918–4924.
[34] Ipek, M., Selvi, I. H., Findik, F., Torkul, O. and Cedimoglu, I. H., 2013. An expert system based material selection approach to manufacturing, *Materials and Design*, Vol. 47, pp. 331–340.
[35] Cicek, K., Celik, M. and Topcu, Y. I., 2010. An integrated decision aid extension to material selection problem, *Materials and Design*, Vol. 31, pp. 4398–4402.
[36] Anojkumar, L., Ilangkumaran, M. and Sasirekha, V., 2014. Comparative analysis of MCDM methods for pipe material selection in sugar industry, *Expert Systems with Applications*, Vol. 41, pp. 2964–2980.
[37] Kan, H. C., Kursuncu, B., Kurbanoglu, C., and Guven, S. Y., 2013. Material selection for the tool holder working under hard milling conditions using different multi criteria decision making methods, *Materials and Design*, Vol. 45, pp. 473–479
[38] Cicek, K. and Celik, M., 2010. Multiple attribute decision-making solution to material selection problem based on modified fuzzy axiomatic design-model selection interface algorithm, *Materials and Design*, Vol. 31, pp. 2129–2133.

[39] Chen, S. J., Hwang C. L., 1991. Fuzzy Multiple Attribute Decision Making: Methods and Applications. Lecture Notes in Economics and Mathematical Systems, No. 375, Springer-Verlag, Berlin, Germany.
[40] Zavadskas, E. K., Kaklauskas, A., Turkis, Z., Tamosaitien, J., 2008. Selection of the effective dwelling house walls by applying attributes values determined at intervals, *Journal of Civil Engineering and Management*, 14, pp. 85–93.
[41] Brauers, W. K. M., 2004. Optimization method for a stakeholder society, *A Revolution in Economic Thinking by Multi-Objective Optimization*, Kluwer Academic Publishers, Boston.
[42] Benayoun, R., Roy, B., Sussman, N., 1966. Manual De Reference Du ProgrammeElectre, Note De Synthese et Formaton, Direction Scientifique SEMA, Paris (1966), No. 25.
[43] Shanian, A., Milani, A. S., Carson, C., Abeyarante, R. C., 2008. A new application of ELECTRE III and revised SIMOS procedure for group material selection under weighting uncertainty, *Knowledge Based Systems*, Vol. 21, pp. 709–720.
[44] Chou, Y. C., Yen, H. Y., Sun, C. C., 2012. An integrate method for performance of women in science and technology based on entropy measure for objective weighting, *Quality & Quantity*, Vol. 48, No. 1, pp. 157–172. http://dx.doi.org/10.1007/s11135-012-9756-6

7 Study of Polar Region Atmospheric Electric Field Impact on Human Beings and the Potential Solution by IPMC

Suman Das
MCKV Institute of Engineering

Srijan Bhattacharya
RCC Institute of Information Technology

Subrata Chattopadhyay
NITTTR

CONTENTS

7.1	Introduction	150
7.2	Atmospheric Electric Field in Polar Regions: Global Electric Circuit	153
7.3	Generation of Air–Earth Current (Maxwell Current)	156
7.4	Diurnal Variation of Fair-Weather Atmospheric Electric Field, Conductivity and Air–Earth Current Density in Polar Regions	156
	7.4.1 Atmospheric Electric Field Measurement	157
	7.4.2 Point Discharge Current: Impact on Polar Vertical Air–Earth Current Density	158
7.5	Environmental Electrostatic Field and Air Ions (Generation of Earth's Vertical Electric Field)	158
7.6	Effects of Human Body Electrostatic Generation in Context with Polar Regions	159
7.7	Energy Harvesting from Human Body Electrostatic Discharge	161
7.8	Investigation into Electroactive Polymer (EAP)-Based Technologies for Human Body Static Nullification in Polar Regions	161
	7.8.1 Selectivity Criteria and Comparison of Different EAP Materials	163

DOI: 10.1201/9781003204664-7

7.8.2　Ionic Polymer–Metal Composites (IPMCs): General
　　　　　　Fabrication Process.. 166
　　　　　　7.8.2.1　Physical Metal Loading 166
　　　　　　7.8.2.2　Casting Method.. 166
　　　　　　7.8.2.3　Hot Pressing Method 166
　　　　　　7.8.2.4　Electrodes Plating Method.. 167
　　　　　　7.8.2.5　Fabrication Method Using Silver Nanopowders 168
　　　7.8.3　Dynamic Modelling of IPMC Actuator...................................... 168
　　　　　　7.8.3.1　Generalized Study on Electrical and Mechanical
　　　　　　　　　　 Parameters of IPMC 168
　　　　　　7.8.3.2　Comparison between Active Sensing, Passive
　　　　　　　　　　 Sensing and Self-Sensing Actuation (SSA)...................170
　　　　　　7.8.3.3　Performance of IPMC Sensors 171
7.9　The Proposed IPMC-Based Methodology (Wearable Electronic
　　　Device Prototype Design)...173
7.10　Discussion and Conclusions..173
References...179

7.1　INTRODUCTION

An extensive array of electromagnetic phenomena is perceived on our earth. Ionosphere, an electromagnetic layer, extends from the earth's surface to an altitude of up to 100–1000 km in the earth's atmosphere. Partially ionized plasma generated by solar radiation such as ultraviolet rays is the dominating atmospheric parameter in that region. Earth's surface is slightly electrically conductive in nature; currents induced due to the changes in the magnetic field in the ionosphere generate electric field through the movement of ion-containing fluids such as groundwater. Naturally, the movement of seawater, which contains vast quantities of ions, also brings about electromagnetic changes [1]. Lightning is one of the most extreme natural phenomena in the field of atmospheric electricity. Static electricity is another form of atmospheric electricity, which is often experienced by wearing a sweater or winter garments in a cold environment or in winter season. The phenomena which include atmospheric electric field that is similar to the earth's geomagnetic field at the global scale during fair-weather days and high-altitude transient luminous events (TLEs) such as sprites and elves that are generated by the interaction of thunderclouds and the ionosphere were reported and witnessed by sailors and pilots since 1990. It is crucial to approach at the altitudes of several kilometres up to 20 km from the ground level where thunderclouds occur. It is not possible to carry out fixed point observation inside thunderclouds while airplanes and weather balloons are used for observation. More importantly, floating observation methods do not provide ground contact, making it difficult to deploy measures for lightning protection [2]. Atmospheric monitoring from sharp mountain peaks has some advantages; for example, the shape of thunderclouds does not change when they pass by such peaks, whereas a collision with a mountain range changes their shape. So, the charging behaviour of the thunderclouds and thereby the nature and the measured value of the atmospheric electric field may diverge. For example, Mt. Fuji's summit, the only high-altitude cone-shaped peak in Japan with an altitude of just under 4 km, reaches the bottoms of thunderclouds, making it possible to place instruments for precise monitoring of the generation of atmospheric electricity in view of global electric circuit. It is also possible to deploy acceptable lightning protection measures by shielding the entire measurement structure in metal to protect it from the frequent lightning strikes at high altitudes. Radiation emitted from minerals containing natural radioisotopes and radiation from outer space known as cosmic rays are the two main sources of naturally occurring radiation on earth. The duration of radiation surges associated with lightning is on the order of milliseconds; the duration of high-energy radiation associated with thundercloud activity is much longer and can last for several minutes. The bottoms of winter thunderclouds are much lower than those of summer thunderclouds. Since radiation is absorbed by the atmosphere, radiation generated by winter thunderclouds, whose bottoms are close to the ground, could be detected at the earth's surface. The atmospheric electric field has a strength of approximately 100 V/m pointing downwards to the earth's surface. Compared to the EMF from a single dry cell battery having a value of 1.5 V, it is obvious that these atmospheric fields are much stronger. This may be inferred from the existence of vertical electric field. The positively charged ionosphere and the negatively charged surface of the earth form the global-scale capacitor. Air ions formed due to the ionization of atmosphere by cosmic rays or natural radioisotopes generate and carry an air-to-earth current of several picoamperes per metre square. The charge stored in the global-scale capacitor would not be discharged by this air-to-earth current. Ionized cloud droplets are also a factor that affects the creation of atmospheric electric field. So, the diurnal variation of atmospheric electric field under fair-weather conditions is taken into consideration for the precise measurement of atmospheric electricity when there are no wind (or very slight wind) and no clouds [3]. Statistical data of diurnal variation in the atmospheric electric field over a certain period of fair-weather days provide a characteristic variation of atmospheric potential gradient. In a graph in which the fair-weather atmospheric electric field is plotted on the vertical axis and the coordinated universal time (UTC) is plotted on the horizontal axis, it is apparent that the atmospheric electric field peaks appear at same value regardless of where in the world the measurement is taken. In other words, the same change is observed around the world, regardless of the local time at each monitoring location. This fact was discovered approximately a century ago based on the measurements taken by the research vessel Carnegie. The nature of variation of the global electric field obeying this condition is called Carnegie curve [1,4,6,7]. The diurnal variation in the fair-weather atmospheric electric field is understood as reflecting the change in overall charges. This understanding of the global atmospheric electric circuit (global circuit theory) explains the recharge mechanism and the reason why the atmospheric electric field does not dissipate even when there is discharge via the air-to-earth current. It was found that in all seasons except for winter, the diurnal variation differs by season and depends on the local time, with the atmospheric electric field increasing as the sun rises and decreasing as the sun sets [8,9]. In winter, the diurnal variation was found to follow the Carnegie curve completely dependent on the UTC [10,11]. Atmospheric electricity shows the evidence of global electricity and global electric circuit, which depend on several physical factors of atmosphere such as the presence of ions, aerosols and radioactivity, which play a vital role in the generation of the global electricity. The ions in the atmosphere attach to several uncharged particles to form a big cluster to form the aerosol. The fast mobility of the positive and negative ions depends upon factors such as temperature, atmospheric density and water vapour content, which are a function of height. The slow variation of electric field in the atmosphere above the open sea water occurs due to the space charge present in the atmosphere due to breaking salt water, which evidences the atmospheric electricity. At sharp mountain peaks, the air–earth current density is much higher in comparison with the flat surface. The latitude effect of air–earth current is stronger than the latitude effect of cosmic radiation. Two highly conductive layers of the solid earth and the lower ionosphere can be assumed as a large spherical capacitor constantly charged by the atmospheric electricity, producing the global electric circuit [1]. The electric field in the human body due to the electrostatic discharge phenomenon along with naturally occurring electric field in case of Polar Regions generates contact currents in the human body, creating health problems due to the imbalance in the body charge. In Polar Regions or at mountain peaks, the vertical atmospheric electric field shows strong existence, which may have some contribution to the generation of human body static electric field. This phenomenon of extremely low frequency (ELF) generation is under investigation in this work. The electrostatic charge accumulation in the human body not only causes health-related problems, but also creates health hazards such as generation of contact current electrostatic shock from spark discharge, and plating out problems where field around the body will enhance the plate-out of negatively charged particles onto the clothes and the exposed skin. But neutral particles will also be attracted, because they will be polarized and the fields, in general, will always be inhomogeneous [12,13]. Okoniewska et al. [14] performed the modelling of interactions of ESD with the human body considering the heterogeneous model of the body. The biomechanical energy due to the accumulation of human body charge should be nullified to avoid health hazards as well as few health-related issues. The proposed electroactive polymer (EAP) materials in the form of some jacket or a garment may be used to nullify the charge accumulation in the human body in Polar Regions [15–18]. EAPs have a high power-to-weight ratio and can produce electrical energy (around 0.8 W) when under mechanical stress under some appropriate solvent or in contact with the moist environment. Different forms of EAPs are available, of which the ionic polymer–metal composite (IPMC) will be under consideration and investigation in this review work [19–22].

The objective of this chapter is to review the different atmospheric parameters responsible for the generation of atmospheric electric field in different environmental conditions and air–earth conductivity in Polar Regions influenced by its diurnal variation. This chapter also investigates the effects of atmospheric electric field in Polar Regions or at sharp mountain peaks. This may imply its contribution to the physiological effects on the health of the people residing at those places. Although a lot of scholarly works provide research contributions in the field of atmospheric electricity and global electric circuit, a complete review of atmospheric electric field and its effects on human health has been missing in the literature under survey to the best of the authors' knowledge. Therefore, the authors felt a need of providing an effective review work considering the effect of Polar Region electric field due to the naturally occurring vertical potential gradient, physiological effects of the electric field on human health and the nullification of static charge accumulated in the body using some advanced technological solutions based on EAPs. This review focuses on the basic principle of global electric circuit in the generation of air–earth current that is considerably high especially in Polar Regions, the naturally occurring body static charge in Polar Regions due to this atmospheric electric field and also the technical aspect of its solution.

This chapter is organized as follows: Section 7.2 provides the ideas of generation of atmospheric electric field influenced by global electric circuit phenomenon. Section 7.3 highlights the generation of air–earth current (Maxwell current), which is the driving force behind the air–earth conductivity. Section 7.4 discusses the diurnal variation of fair-weather vertical atmospheric electric field. Section 7.5 provides a summary of naturally occurring environmental static electric field. Section 7.6 provides different case studies of biological effects of human body static generation, especially emphasizing the polar atmospheric conditions. Section 7.7 discusses different energy harvesting processes based on human body static generation. Sections 7.8 and 7.9 investigate the proposed EAP-based technological solutions to nullifying the body static charge generated in Polar Regions, and Sections 7.10 and 7.11 provide an overall discussion on the whole review work and the concluding remarks, respectively.

7.2 ATMOSPHERIC ELECTRIC FIELD IN POLAR REGIONS: GLOBAL ELECTRIC CIRCUIT

The global electric circuit in view of the atmospheric global electric current which flows from ionosphere via troposphere due to thunderclouds makes the return path to the earth's surface at the fair-weather days completing the global electric circuit (Figure 7.1). Atmospheric aerosols also have an impact on the fair-weather atmospheric potential gradient. So, the earth's changing climate phenomena are monitored by the global atmospheric electric circuit Rycroft et al. [23]. In comparison with the air, everything on the earth is considered as good conductors. Human body is also a very good ionic conductor; therefore, the ever existing atmospheric electric field is not experienced always. According to Lord Kelvin, the capacitive effect exists between the ionosphere (positive plate) and the earth's

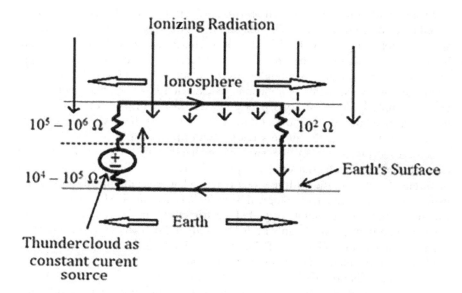

FIGURE 7.1 The global electric circuit generating electrical environment between the ionosphere and the earth's surface through the atmosphere. The sum of thunderstorms around the world symbolizes the current generator; the charging and discharging occur through different resistors above and below the thunderclouds.

surface (negative plate), establishing a potential difference of 260 kV and thereby producing an atmospheric discharge current of 1 kA in the form of thunderstorm [24, 25]. With respect to the thundercloud, the upward flowing current towards ionosphere is called the Wilson current when the downward flowing current is termed the air–earth current. The atmospheric conductivity depends on the ionic concentration of the atmosphere over the earth's surface. The source of ionization is the collision between the galactic cosmic rays with the neutral molecules present in the atmosphere. The nitrogen and oxygen molecules when combined with water molecules form the aerosols. The formation of aerosols reduces the air conductivity. The near-surface air – current conductivity is of the order of 10^{-14} Siemens per metre (S/m), whereas at 100 km altitude, the conductivity increases 11 times to become 10^{-3} S/m [50–55].

Ions present in the atmosphere attach with several uncharged particles to form a big cluster of aerosols. The fast mobility of the positive and negative ions depends upon the factors such as temperature, atmospheric density and water vapour content, which are the functions of altitude [28–31]. At sharp mountain peaks, the air–earth current density is much higher and prominent in comparison with the flat surface. The electrical structure of the earth–ionosphere system can be modelled as a leaky spherical capacitor. The surface of the earth and the highly conductive ionosphere (above 50 km) form the inner

Polar Atmospheric Electric Field & IPMCs

and outer shells, with the atmosphere functioning as a leaky dielectric. Singh et al. [35] performed a review on the electrical environments of the earth's atmosphere with different sources of electric fields in different regions and their interconnection through electrodynamic coupling. The earth's surface is assumed to consist of negative charge, whereas the equal and opposite positive charge is distributed all over the atmosphere above the surface. In the lower atmosphere which comprises the troposphere, stratosphere and mesosphere, the thunderstorm activity produces vertical electric fields, which provides the main source of electric fields in those regions. According to Freier's thunderstorm model, the atmosphere between the earth and the ionosphere is subdivided into three separate regions, such as the region below the negative layer of the charged thunderstorm cloud, the region between the bottom and top of the thunderstorm cloud and the region above the thunderstorm cloud. The atmospheric electric field distribution is sustained by thunderstorms in the form of thunderclouds. A thundercloud normally has a negative base and a positive top transmitting negative charge to ground. Due to the presence of strong atmosphere, the electric field around the thunderclouds resulting in positive air–earth current completes the global electric circuit. The resistance 'r' represents the parallel resistance of the air columns during fair-weather days (Figure 7.2). I_0 represents the generator effect of all simultaneously active thunder systems, and R_1 and R_3 represent the resistances of the air columns above and below these systems, respectively, whereas R_2 represents the internal source resistance of the current generator representing thunderclouds.

The medium through which the atmospheric air–earth current will flow must contain air ions. Molecular clusters of charge carriers or atmospheric ions are the main charge carriers. Neutral oxygen or nitrogen molecules when energized lose an electron to form a single positively charged elementary ion. The electron when combines with the oxygen molecule forms a negatively charged elementary ion within a microsecond. Atmospheric ions include hydroxonium (H_3O+) and charged nitrogen oxides, which have a very short lifespan. In

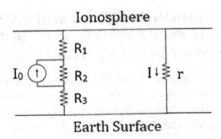

FIGURE 7.2 Electrical equivalent global electric circuit, implying the global thunderstorm current generator by 'I_0' and the fair-weather current by 'I'.

polluted city air, the average lifespan is in the order of 10–20 seconds, whereas in polar air, having poor aerosol content, the lifetime may be as long as 300–400 seconds [32–34] and [36–38].

7.3 GENERATION OF AIR–EARTH CURRENT (MAXWELL CURRENT)

In Polar Regions, the local influence on the atmospheric electric field is very less, so by measuring the total Maxwell current flowing to the surface during fair-weather days, the global atmospheric electric current can be accurately measured. The Maxwell current is the atmospheric phenomenon involved in the generation of air–earth current (convection current) [2]. The electric field changes more than the change in air–earth current due to the local change in atmospheric conductivity. A local diurnal variation does not exist in the Polar Regions, and hence, the measurement of field in these regions could represent the global variation [4]. The time variation in thunderstorm electric fields is considered as the total Maxwell current density that varies slowly. The total Maxwell current

$$J_M = J_E + J_C + J_L + \partial D/\partial t \qquad (7.1)$$

In Eq. (7.1), J_E is the field-dependent current which includes the point discharge current, J_C is the convection current produced by the mechanical transport of charge such as by the air motion or by precipitation, J_L is the lightning current that represents the discontinuous transfer of charge in both space and time and hence occurs impulsively, and $(\partial D/\partial t)$ is the displacement current. The thundercloud is assumed as a constant current source having positive charge at the top and negative charge at the bottom of the thundercloud Singh et al. [35]. Ruhnke [28] introduced a new method based on area averaging of surface measurements of the atmospheric electric current density, reducing the local disturbance effects.

7.4 DIURNAL VARIATION OF FAIR-WEATHER ATMOSPHERIC ELECTRIC FIELD, CONDUCTIVITY AND AIR–EARTH CURRENT DENSITY IN POLAR REGIONS

Siingh et al. [41] explored various atmospheric electric parameters and the diurnal variation of fair-weather electric fields in fair-weather atmospheric conditions (wind speed of ≤10 m/s) at a relatively pollution-free sub-auroral station Maitri (70°75/S, 11°75/E), during the 12 fair-weather days of January and February 2005. According to this study, the influence of small and large ions is significant on the air–earth current density in the fair-weather days (about 1–6 pA/m² current flows vertically between the ionosphere and the earth's surface). The sources of error in the measurement of global electric field are influenced by the presence of local generations. The air–earth conductivity will be a lesser value due to the presence of atmospheric aerosols. The diurnal variations of atmospheric parameters are seasonal variations at the tropical regions [6]. Bennett et al. [44,45] and [48] investigated the atmospheric electrical parameters at different weather conditions such as fair-weather conditions, disturbed weather and the presence of fog to vary the potential gradient. Snow shower and rain shower are the influencing factors behind the variation in the potential gradient.

7.4.1 Atmospheric Electric Field Measurement

Willett et al. [49] reported on the contact potential and surface charge effect on atmospheric electricity. In the lower atmosphere, the polar conductivity was measured and the value was around 1.8×10^{-10} ℧/m. The capacitive probe method explored the effects of surface charge on the variation of electrical properties of the metal surface [5]. Various atmospheric electrical parameters such as vertical potential gradient (E), vertical air–earth current density (J_z) and atmospheric electrical conductivity (σ) in fair-weather conditions are measured in the Northeast India campaign [7]. The average vertical potential gradient was found to be 108 V/m and the air–earth current density was 1.85 pA/m², maintaining the positive correlation between E and J_z. The properties of dust storm during electrification were studied and measured for different atmospheric electrical parameters such as atmospheric potential gradient, space charge and wind speed properties, temperature and relative humidity and size distribution of dust particles at ground level to study the influence of the dust particles on the atmospheric electric field. The daily measurement of electric conductivity is more or less invariant at Maitri as well as over open ocean. Fog cover reduces the electric conductivity due to the presence of small ions, which are fog particles that reduce the mobility and thus the electrical conductivity of the atmosphere. The electrical conductivity over the Indian Ocean during eight fair-weather days was reported. They observed that the conductivity values during night-time are comparatively higher than those during the daytime. The average value of the total conductivity for the eight fair-weather days along the cruise is 1.74 S/m. The local influences on the vertical component of the geoelectric field at the Davis station, Antarctica, based on the 'cumulation of consecutive difference method' averaged over year-wise collected data of 'fair-weather' conditions were investigated by Burns et al. [59]. The local influence range between 03 UT and 10 UT evidences the magnetospheric influence on the geoelectric field. In the Polar Regions, the interaction between the solar wind and the earth's magnetic field leads to the establishment of a variable electric potential difference of 20–150 kV between the dawn and dusk sectors of the polar cap. Blakeslee et al. [41] established that electrified storms are the sources of the fair-weather electric field. The atmospheric electricity is an ever persisting phenomenon on the earth's surface, but is much predominant during fair-weather days. The atmosphere on the fair-weather days behaves like a loose dielectric medium, and the air conduction current density is around 1–6 pA/m² [56–58] and [60].

7.4.2 Point Discharge Current: Impact on Polar Vertical Air–Earth Current Density

The interchange of electricity between the earth and the atmosphere due to point discharge phenomena plays an important role in the generation of atmospheric electric field. Point discharge phenomena were first observed at Kew Observatory, which illustrates the atmospheric effect on the point discharge phenomena [39]. The relation between the potential gradient and point discharge current was found above a critical minimum value of the field current, which is roughly proportional to the square of the gradient. Hutchinson [27] reported the measurement data of point discharge current, which increases with the square of the field near the earth's surface. In case of corona discharge, the point current will increase linearly with field. The point discharge phenomena become much precise at high latitudes. The presence of space charge over the surface of earth at different locations attributes to the local point discharge and thereby the vertical electric field. An empirical formula explains the relationship between the point discharge current and other physical atmospheric parameters: $i = a\ (FZ - MZ)$, where i is the downward directed current in pA, F the downward directed field in v/cm, M the minimum field for the onset of point discharge, and 'a' the constant depending on the sign of the current. Markson et al. [61,62] investigated the potential gradient at mountain peaks, which results Andes glow due to corona discharge at the fair-weather climatic conditions. Due to eddy diffusion, a strong potential gradient present at the sharp peaks of mountains results in ionic point discharge also known as brush discharge. The large magnitude of electric fields due to point discharge evidences the atmospheric potential gradient ranging between 1150 and 3100 V/m (the electric field strength) at the peaks and ridges of high mountains during fair-weather conditions.

7.5 ENVIRONMENTAL ELECTROSTATIC FIELD AND AIR IONS (GENERATION OF EARTH'S VERTICAL ELECTRIC FIELD)

Electrostatic discharge (ESD) events are related to static electricity that creates some hazardous effects in view of human health as well as damage to sophisticated electronic systems. ESD phenomena are assumed to be the result of human activities. The atmospheric electricity which is due to natural phenomena is much analogous to static electricity processes. If there is a thundercloud overhead, the field is usually reversed and runs easily into the tens of kV/m. If a horizontal metal plate exposed to the free atmosphere is connected to ground through a sensitive ammeter, it would measure a current of about 3×10^{-12} A/m². A value of 3 pA/m² is not much, but when it is taken for the earth as a whole, the current amounts to about 1500 A [63]. According to the study reported by Charry [64], positive ions have much adverse effects on human physiological aspects in comparison with negative ions. This trend suggests that air ion exposure has an effect on humans and animals. Electrostatic discharges (ESDs) produced in the human tissue due to biophysical interactions with the environment are evaluated by comparing the fields.

7.6 EFFECTS OF HUMAN BODY ELECTROSTATIC GENERATION IN CONTEXT WITH POLAR REGIONS

One of the ways of human body static charge generation is walking across an insulated floor covering. The contact and friction between the shoe soles and the floor cause a charge separation for each step. This charge makes the voltage of the human body capacitance increase until the unavoidable leakage current balances the charging current. This effect might well raise the person's voltage to substantial levels, with the net charge remaining zero. Charges get separated by rubbing fabric materials such as cloths, but the voltage of the person will not increase, since equally large opposite charges are in principle located on the person. But when the winter garment, especially that woven by wools, carrying negative charge is removed, the positive charge from the human body itself provides a positive voltage. Incidentally, removing the woollen garments (sweater, jacket, etc.) is not a sign of charging, but rather it's a discharge (and it's not a spark, but a brush discharge) [70–74]. Spark discharge phenomenon is well known, but there are no well-defined ranges for what level of body voltage will result in discharges that can be felt. Petri et al. [75] studied through a systematic review process the biological effects due to the exposure to static electric fields on humans and vertebrates. The results show that humans and animals can possess a broad range of static electric field strength. Static electric fields (EFs) arise naturally in the environments such as storm clouds in case of atmospheric electric field generation or human body static generation through triboelectric charge separation on clothing or some other ways such as touching some charged or uncharged objects; the body charge balance gets disturbed, which results body static electricity generation. The atmospheric static EF (field strength ranging between 0.1 and 0.3 kV/m at ground level) is generated between the positively loaded ionosphere and the negative ground, and it depends on the seasonal variation and the quality of the atmospheric parameters such as temperature, aerosol content and air–earth conductivity. There are various sources of generation of human body static electric charge, such as natural atmospheric potential gradient, the vicinity of HVDC transmission lines, cathode-ray tube displays, trams and urban railways. For example, ±600 kV HVDC transmission lines produce around 35 kV/m static EF around them. The static EF generated from an external source is attenuated by the factor 10^{-12} while entering into the human body. So, with the exposure of human body to static EF, the charges are redistributed on the body surface, which sometimes produces a large surface charge density resulting in spark discharges (microshocks) [76,77]. In the simultaneous exposure of air ions on the human body specially observed in the Polar Regions, the sensitivity of electric field detection in human body increases, permitting subjects to detect the EF at lower field strengths. The threshold values of EF detection sensitivity affects only when ion current density values are greater than 60 nA/m^2. But the air–earth current density in the fair-weather days influenced by small and the large ions is about 1–6 pA/m^2 and flows vertically between the ionosphere and the earth's surface in the Polar Regions [35]. Dawson et al. [78] described the effects of electric field on the human body due to ESDs. The hypothesis on ESD leads to the idea that imperceptible contact currents in the human body may cause health hazards, especially in the electric utility workers. The human body static charge elimination technologies based on body electrostatic potential are influenced by different types of physical movements. Grounding bracelet in the form of anti-static wrist strap can reduce human electrostatic potential in any state of movement. Charge may be accumulated in human body due to various human movements such as walking, running and touching some charged objects. Electrostatic induction and space charge adsorption are the two ways of accumulation of body static, resulting in the unbalanced body charge distribution. Body static may cause several hazardous effects such as fire disaster and contact hazards. Anti-static shoes, anti-static suit, anti-static wrist strap and grounding wrist strap are applied in this study to ground the human body to eliminate the ESD hazards [79–82]. The imbalance in the potential between the human body and the object in contact results in contact current which gets grounded through human body, resulting in hazardous effects to humans. Human body is an anisotropic system having different electric resistivities in different limbs. Several microamperes (nominal value around 0.1 mA body-to-ground) of contact current can produce a considerable amount of persistent electric potential gradient (mV/m) within the body. Repacholi et al. [86] reported on the biological effects and the related health hazards of ambient or environmental static and extremely low frequency (ELF) electric and magnetic fields (0–300 Hz) [83–85]. The working groups concluded that, although health hazards exist from exposure to ELF fields of high strength, the literature does not establish that health hazards are associated with exposure to fields of low strength, including environmental levels. An electrically grounded person touching an insulated metal object will draw current from the object and experience an electric shock or spark discharge. Similarly, a person walking across a carpet in a dry atmosphere becomes charged and discharges on touching a grounded metallic object. Aside from the process of charging and discharging, no other direct action of static EFs on living systems is known. Fields of up to 600 V/m had no apparent effect on circadian rhythm in humans, and no consistent effects have been reported after exposures to fields of up to 12 kV/m. No adverse health effects are known to occur from exposure to static EFs [87–89]. Nowadays, most of the time human body remains insulated from the ground, resulting in the storage of charge and thereby the production of human body capacitance (around 100 pF) which generates body potential (volts). By the study and validation of experimental results, Kent Chamberlin et al. [90] concluded that the low earthing current between human body and ground doesn't carry any physiological information, but can detect only the movement of human body which generates static charge to flow through the ground [64,91]. Different human and mechanical activities such as docking spacecraft and implementation of external payloads can also produce an electrostatic environment at the International Space Station (ISS) [92]. Pandey et al. [93] investigated the fundamental relationship between the geometry of the material and the electrostatic charge storing capacity. The findings show that when a charged material becomes more compact in shape, the amount of charge decreases, and vice versa, following the principle of reversibility. Electrostatic build-up and its uncontrolled discharge can have serious consequences for people – with both direct and indirect ramifications for the person affected. By way of example that describes two scenarios, the difference between protective anti-static and ESD clothing is illustrated [94–96]. Two bodies coming into contact – for example an upper arm rubbing against a torso as two people walk past each other – causes an electrostatic build-up. Anti-static, on the other hand, occurs in potentially explosive environments in the chemical industry; fuel depots; refineries; silos; and conveying, mixing and milling plants [97].

7.7 ENERGY HARVESTING FROM HUMAN BODY ELECTROSTATIC DISCHARGE

The energy released from the human body can be harvested for powering electronic devices. The mechanical efficiency of the human body is about 15–30%. The total sensible heat that is released into the atmosphere by a person walking at natural speed is approximately 100 W, but the conversion factor is 2.15%; therefore, the maximum power available during walking would be approximately 2 W. A jacket or a garment is used to cover the body surface for energy harvesting. EAP materials which have a high power-to-weight ratio can produce electrical energy (around 0.8 W) under mechanical stress, but they have a low efficiency (compared to magnetic machines), i.e., of 2 W during normal walking. EAPs can produce greater strain compared with piezoelectric materials, so more energy can be harvested. EAPs are much light in weight and easier to shape than magnetic materials. Raziel Riemer and Amir Shapiro [98] predicted that EAPs are a good alternative to piezoelectric materials for biomechanical energy harvesting, but experimental results or data for validation of that conclusion were absent in their study [99–101]. The electric field in the human body due to ESD phenomena can generate contact currents in the human body that may cause health problems. An ESD through a human body has peak currents of the order of a few to several tens of amperes and a relatively short duration of the order of 100 ns, which produces microshocks during contact [78].

7.8 INVESTIGATION INTO ELECTROACTIVE POLYMER (EAP)-BASED TECHNOLOGIES FOR HUMAN BODY STATIC NULLIFICATION IN POLAR REGIONS

Polymers are inherently lightweight, mechanically flexible and easy for mass production and processing. Electrical excitation causes elastic deformation in polymers. Some of the EAP materials exhibit the reverse effect of converting mechanical strain to an electrical signal; these phenomena make them useful for sensors and energy harvesting applications. According to the working principle,

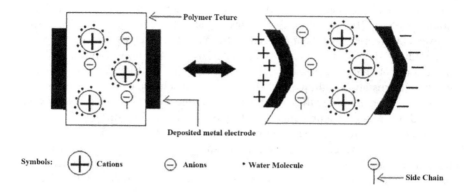

FIGURE 7.3 The typical ion exchange in IPMC strip without and with applied voltage representing the actuation phenomenon. Water leakage through the porous Pt electrode reduces the electromechanical conversion efficiency. (Courtesy of Bhattacharya et al.'s Mechanics Based Design of Structures and Machines [210].)

EAPs are categorized in two groups: ionic EAPs and electronic (also known as field-activated) EAPs. The activation mechanism of ionic EAPs depends on the transport or diffusion of ions. Two electrodes and an electrolyte are the constituting materials of this type of EAPs. IPMCs, conductive polymers, carbon nanotubes and ionic polymer gels follow the working principle of ionic EAPs (Figure 7.3). Large bending under low activation voltage (1–2 V) is the main advantage.

The limitation of such types of EAPs is the requirement of sustained electrolyte wetness that makes their efficiency relatively lower in comparison with other forms of EAPs. The ionic diffusion also makes the actuation speed as well as the processing speed slower. An electronic EAP creates an electric field between the electrodes on a film-shaped polymer material. The piezoelectric polymers, such as PVDF, exhibit a linear relation between the electric field and the generated strain, but produce a low strain. For activation, high voltage levels (>10 V/mm) are required, which may be close to the breakdown level of the dielectric constant of the polymer. As the actuation is independent of diffusion of charge carriers, these types of EAP materials possess fast response speed (milliseconds). Examples of these materials include electrostrictive, electrostatic, piezoelectric and ferroelectric materials [15–18].

Shahinpoor et al. [105] presented the first review among the four about the IPMC, where it is said that the IPMC is a perfluorinated polymer membrane which is sandwiched with electroplated platinum (Pt) on both sides and polymers having ion exchange capability. The IPMC material belongs to the ionic EAP class where the actuation is caused by ion diffusion. IPMC can be used as actuators and sensors. These polymer films are Nafion or Flemion. In Nafion-based IPMCs, the slow relaxation is towards the cathode, whereas in Flemion-based IPMCs, the slow relaxation continues the initial fast motion towards the anode. The ion exchange polymers refer to polymer design to selectively exchange ions of a single chain (cations or anions) with their own initial ions. They also said that the manufacturing process follows two major steps: First, the initial compositing process and, second, the surface electroding process. Polymers are treated with an ionic salt solution of platinum ammine (with free ion, Pt(NH$_3$)HCl) of a metal and then chemically reduced to yield IPMCs. The polymer is soaked in the salt solution, which makes the platinum-containing cations diffuse through the ion exchange process. The reducing agents such as LiBH$_4$ (lithium borohydride) or NaBH$_4$ (sodium borohydride or sodium tetrahydridoborate) are used to metalize the polymer. The bending force of IPMC is generated by the effective redistribution of hydrated ions and water; basically, it is an ion-induced hydraulic actuation phenomenon. This bending force is dependent of the electric field and the length of the IPMC. The advantage of the IPMC is that it can be used in both air and water.

When potential difference is applied across the terminals, ion exchange caused through polymer membrane and microfilmic platinum is diffused in the polymer substrate.

$$\text{LiBH}_4 + 4\left[\text{Pt}(\text{NH}_3)_4\right]^{2+} + 8\text{OH}^- \Rightarrow 4\text{Pt}^0 + 16\text{NH}_3 + \text{LiBO}_2 + 6\text{H}_2\text{O} \quad (7.2)$$

The actuation of IPMC strip is determined by the electric field applied to the terminals [105]. Although the capacitance of IPMC is prominent at low voltage, with increasing potential difference, the capacitance decreases by 0.20 times of its initial value [133]. The IPMC yields charge reorganization with one part of the supplied current, which results in a stress inside the IPMC with two components: blocked force and free deflection [208]. The IPMC is applicable for air or water environment. The limitation in its performance depends on the ambient condition specially affect in dry atmosphere. Shahinpoor and Kim [106,107] presented a description of various modelling and simulation techniques and the associated experimental results in connection with IPMCs as soft biomimetic sensors, actuators, transducers and artificial muscles. Malone and Lipson [108] made a literature survey on electromechanically active materials, conducting polymers and proton exchange membrane materials. They tested IPMCs, and when they get dry, there is a reduction in the operating life. The silicone-encapsulated Nafion-based device tip elevated to about 1 mm in 50 seconds in response to 5 V. Leary and Cohen [110] of Jet Propulsion Laboratory, CA, used a Nafion-based IPMC sample of dimension 30 mm × 5 mm × 0.2 mm with platinum electrodes and mobile Na+ counterions to explain the electrical impedance of the IPMC. The application of a cosine wave of 7 V and 0.03 Hz to the sample results in a nonlinear hysteresis behaviour. Authors concluded that at large bending actuation occurs at low voltage level also said at 4 V current at steady state is 0.6 A.

7.8.1 Selectivity Criteria and Comparison of Different EAP Materials

The following tables present different forms of EAP materials and their working principles and some characteristic properties through which they are selected for some specified applications [15,21,117–121] (Tables 7.1 and 7.2).

TABLE 7.1
Comparison between Different Field-Activated EAP Materials and Ionic EAPs [18]

EAP Material Types	Principle	Reported Materials
Field-Activated EAPs		
Ferroelectric polymers	These polymers exhibit spontaneous polarization that can be switched by external electric fields. They can exhibit piezoelectric response when poled and electrostriction are in nonpolar phase.	Electron-radiated poly(vinylidene fluoride–trifluoroethylene)
	Recent introduction of defects in PVDF–TrFE copolymer crystalline structure by electron irradiation or copolymerizing with a third bulky monomer dramatically increased the induced strain.	PVDF–TrFE-based terpolymers
Dielectric EAPs or electrostatically stricted polymers	Coulomb forces between the electrodes squeeze the material, causing it to expand in the plane of the electrodes. When the stiffness is low, a thin film can be shown to stretch more than 100%.	Silicone Polyurethane
Electrostrictive graft elastomers	Electric field causes molecular alignment of the pendant group made of graft crystalline elastomers attached to the backbone.	Modified copolymer – PVDF–TrFE
Ionic EAPs		
Ionic gels	Application of voltage causes movement of hydrogen ions in or out of the gel. The effect is a simulation of the chemical analogue of reaction with acid and alkaline.	Poly(vinyl alcohol) gel with dimethyl sulphoxide Poly(acrylonitrile) with conductive fibres
Ionic polymer–metal composites (IPMCs)	The base ionomers provide channels for cations to move in a fixed network of negative ions on interconnected clusters. Mobile cations from the anode are responsible for the bending actuation.	Base ionomers: Nafion® (perfluorosulphonate, made by DuPont) Flemion® (perfluorocarboxylate, made by Asahi Glass, Japan) Cations: tetra-n-butylammonium, Li+ and Na+ Metals: Pt and gold
Conductive polymers (CPs)	Materials that swell in response to an applied voltage as a result of oxidation or reduction, depending on the polarity causing insertion or de-insertion of (possibly solvated) ions.	Polypyrrole, poly(ethylene dioxythiophene), poly(p-phenylene vinylene)s, polyaniline and polythiophenes

(Continued)

TABLE 7.1 (*Continued*)
Comparison between Different Field-Activated EAP Materials and Ionic EAPs [18]

EAP Material Types	Principle	Reported Materials
Carbon nanotubes (CNTs)	The carbon–carbon bond of nanotubes (CNTs) suspended in an electrolyte changes in length as a result of charge injection that affects the ionic charge balance between the CNTs and the electrolyte.	Single- and multiwalled carbon nanotubes

TABLE 7.2
Characterizing Properties for EAP Materials [150]

Measurement	Properties	Significance
Mechanical	Tensile strength [Pa]	Mechanical strength of the actuator material
	Stiffness [Pa]	Required to calculate blocking stress, mechanical energy density and mechanical loss factor/bandwidth
	Thermal expansion coefficient [ppm/C]	Affects the thermal compatibility and residual stress
Electrical	Maximum voltage [V]	Necessary to determine limits of safe operation
	Impedance spectra [ohm and phase angle]	Provide both resistance and capacitance data. Used to calculate the electrical energy density, electrical relaxation/dissipation and equivalent circuit
	Nonlinear current [A]	Used in the calculation of electrical energy density; quantify nonlinear responses/driving limitations
	Sheet resistance [ohms per square]	Used for quality assurance
Microstructure	Thickness (electrode and EAP), internal structure, uniformity, anisotropy and hosted defects	These are features that will require establishing standards to assure the quality of the material
Electroactive properties — Strain	Electrically induced strain [%] or displacement [cm]	Used in the calculation of 'blocking stress' and mechanical energy density
Stress	Electrically induced force [N] or mechanically induced charge [C]	Electrically induced force/torque or stress-induced charge
Stiffness	Stress–strain curve	Voltage-controlled stiffness
Environmental behaviour	Operation at various temperature, humidity and pressure conditions	Determine material limitations at various conditions

7.8.2 Ionic Polymer–Metal Composites (IPMCs): General Fabrication Process

The general fabrication process of IPMCs is discussed with a detailed demonstration of all the steps followed to fabricate an IPMC sensor or IPMC actuator.

7.8.2.1 Physical Metal Loading

This is a novel fabrication process for manufacturing IPMCs with physically loaded electrodes. The fabrication process consists of two steps. First, physically load a conductive primary powder into the ionic polymer, forming a dispersed particulate layer. Second, it is further secured with smaller secondary particles via chemical plating, similar to the conventional methods (using reducing agents). Furthermore, an electroplating process can be applied to integrate the entire conductive phase intact, which serves as an effective electrode Bhandari et al. [122].

7.8.2.2 Casting Method

The thickness of the commercially available Nafion® film varies from 50 to 180 μm, and the performance of the IPMC varies with its thickness, such as bending stiffness and electric field. Hence, to achieve a Nafion film of desired thickness other than the commercialized thickness, the casting method is of great use. The overall casting process consists of three steps: (1) mixing, (2) stirring and sonicating and (3) thermal treatment. Thermal treatment increases the mechanical stiffness of the cast Nafion® film by increasing the strength of the molecules' bonding structure. Finally, the cast Nafion® film has to be boiled in the hydrogen peroxide solution in the temperature between 75 and 100°C for 1 hour and then boiled in the DI water for 1 hour. The casting method is required to achieve thick IPMC actuators with Nafion [122,123].

7.8.2.3 Hot Pressing Method

The hot pressing method, as shown in Figure 7.4, uses a hot pressing system to make several thin Nafion® films adhere together, which enhances the bending stiffness, the force performance and the reproducibility. The main advantages of this method are its simplicity, repeatability and easy control of the thickness. The fabrication of IPMC actuators by this process is described as follows. Tailor the Nafion® films with proper dimensions, rinse with acetone and stack in the mould with polyimide films. The mould is then placed between the preheated presses at 180°C without pressure for 20 minutes and is pressed for 10 minutes at 50 MPa and 180°C. Cool the Nafion® films to ambient temperature in the air, and boil the film in 3wt% sulphuric acid at 70°C for 1 hour, in 10wt% hydrogen peroxide at 70°C for 1 hour and in DI water for 0.5 hour. Immerse the film in an aqueous solution of platinum ammine complex ([Pt $(NH_3)_4Cl_2$]) for 1–2 days to allow Pt ions to be absorbed in the stacked film. After rinsing the film with DI water, place it in 500 mL of stirring water at 40°C and add 5 mL of 5wt% sodium

Polar Atmospheric Electric Field & IPMCs

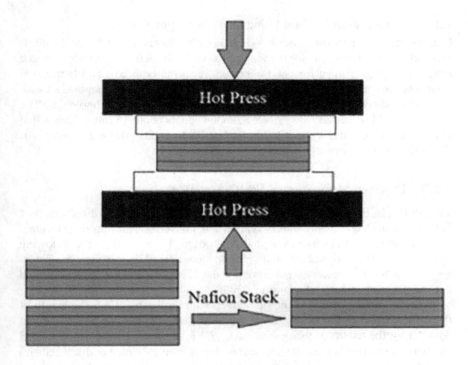

FIGURE 7.4 Fabrication by stacking and hot pressing.

borohydride (NaBH$_4$) solution every 30 minutes as the temperature gradually increases up to 60°C [117,122,123].

7.8.2.4 Electrodes Plating Method

There are several fabrication methods including sputtering, taping with conductive films and chemical reduction. However, the IPMC is generally fabricated using the chemical reduction method due to the low cost and good surface toughness obtained using this method. The adhesion between the polymer and the electrode is one of the main problems during the IPMC fabrication. The electroless method is used with other processes such as hot embossing, electroplating and polymer coating to increase the adhesion and decrease the cost and fabrication time [125,126]. Palladium, platinum and gold are commonly used as the electrode materials due to their corrosion resistance and high conductivity. Electrode deposition on dry and solvated membranes, using a direct assembly process, has also been considered. However, electrode deposition is the most commonly used manufacturing process because this process results in mechanically and chemically stable electrodes. Kim et al. [153] reported that concentration and size distribution of platinum particles have a direct effect on the performance of IPMCs.

7.8.2.5 Fabrication Method Using Silver Nanopowders

Chung et al. [125] proposed microfabrication technologies that integrate the physical and chemical methods for the fabrication of IPMC actuators without surface roughening pretreatment for good adhesion, reducing both cost and fabrication time. The basic principle of this method is the casting of silver nanopowders with Nafion® followed by embossing and silver electroless plating technologies. The thickness of films casted using silver nanopowders is about 15 μm, whereas that of the platinum casting is about 1–20 μm. The IPMC used in this experiment is of dimension 2 cm × 5 mm × 0.23 mm.

7.8.3 Dynamic Modelling of IPMC Actuator

The IPMC electromechanical conversions occur in the transverse direction: they bend when an electrical stimulus is applied and, conversely, they produce an electrical signal when a mechanical signal is imposed (Figure 7.5). In the following paragraphs, the configuration and relevant geometrical parameters are introduced. All the parameters characterizing the IPMC membrane will be scaled with respect to geometrical quantities [209].

Here, 'F' is the force, u is the tip deflection of the beam, 'L_c' is the length of the clamped part of the beam, 'L_t' is the total free length of the beam (without considering the length of the pinned part), 'L_s' is the point where the developed mechanical reaction is desired (actuator) or the mechanical (force or displacement) stimulus is applied, and 'w' and 't' are the dimensions of the beam cross section.

7.8.3.1 Generalized Study on Electrical and Mechanical Parameters of IPMC

In order to assess the electrical properties of the IPMC, the equivalent electric circuit model is required to be developed. It is to note that the IPMC is nearly resistive (>50 Ω) in the high-frequency range and fairly capacitive (>100 μF) in the low-frequency range. Based upon the above findings, the simplified equivalent electric circuit of a typical IPMC is considered. In this approach, each single unit circuit (i) is assumed to be connected in a series of arbitrary

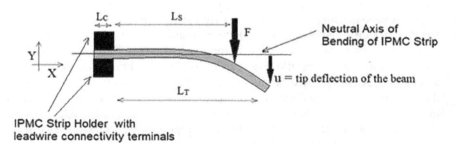

FIGURE 7.5 Mechanical configuration of IPMC beam with system parameters.

surface resistance (R_{ss}). This approach is based upon the experimental observation of the considerable surface electrode resistance [105].

Cohen [149] showed the use of EAPs as artificial muscles. A comparison is made between EAPs, EACs (electroactive ceramics) and SMAs (shape-memory alloys), and it has been seen that the properties of EAP offer superior capability. Yoseph Bar-Cohen et al. [150] studied the characteristics of IPMC of dimension 7.7 mm (length) × 3.3 mm (width) × 0.196 mm (thickness); they studied the response in water at 6.5 mm from the sample tip, where a resonance was observed at 113 Hz in water and 272 Hz in air. Zheng Chen et al. [151] developed a microfinger with IPMC/PVDF [127–129]. PVDF film (40 μm thick) is bonded to an IPMC sample of 50 mm (length) × 5 mm (width) × 0.3 mm (thickness). Paquette et al. [152] described the behaviour of IPMC in multilayer configuration. Three identical, sandwiched squares of IPMC of dimension 1 mm (length) × 1 mm (width) × 0.2 mm (thickness) are used for this. Here, it is seen that the response time is much slower than that of a typical IPMC. Kim et al. [153] measured the stiffness, displacement and force of an IPMC (dimension 30mm × 4 mm × 0.4/1.2 mm). The input given to the IPMCs is a 2–5 V_{pp} AC sine wave with 0.5 Hz frequency. The stiffness of the IPMC is measured with the governing equation $\delta = (PL^3)/(3EI)$, where δ is the measured tip displacement of the actuator, P is the force applied to the actuator, L is the length of the actuator, E is Young's modulus, and I is the moment of inertia. The stiffness of samples is proportional to the third power of the thickness $EI = E[(bh^3)/12]$, where b is the width and h is the thickness of the IPMC actuator. It is seen from the graph that the stiffness and force increase with increasing thickness of IPMC, but the displacement increases with decreasing thickness. The Young's modulus of IPMC with Pt electrode is given by 83.5 MPa. Nasser et al. [154] verified the proposed micromechanical model with Nemat-Nasser's model by applying 1 V to IPMC increase 0–10–20 seconds. The total transported charge by cation during actuation was obtained by the time integration of the measured current. The current gradually increases to 9.2 mA at 10.0 seconds and then decreases to 0.26 mA. Nasser [155] tested IPMC with ethylene glycol (EG) other than water. The backbone of the perfluorinated ionomer is Nafion or Flemion. This membrane is plated with noble metals such as platinum and gold. The IPMC in K+ and Na+ form is tested. The Nafion-based IPMC in K+ form shows faster actuation than the Nafion-based IPMC in Na+ form (see Figure 2.1). From this experiment, it can be concluded that the EG solvent is better only for the longer actuation in air and higher volume uptake. Other than that, the water solvent is good for IPMC for faster actuation, overall capacitance and greater stiffness.

Zhe Lu et al. [156] surveyed the critical issues and the available solutions related to force control in micromanipulation. In micromanipulation, the size of the manipulated object is usually much less than 1 mm in a single dimension, in which case gravitational and inertial forces are no longer dominant. The paper focused on techniques for dealing with adhesion forces [134–139]. Adhesion forces may arise when an object with size less than 1 mm in a single dimension is in contact with (or in close proximity to) another object. Adhesion forces can be reduced by altering the physical characteristics of the object and its environment, such as by reducing the electrostatic forces, van der Waals forces and surface tension forces. To accomplish the end effector easy movement or release an object, instantaneous acceleration of the end effector is reduced by reducing the adhesive force on the manipulator. There are many methods for force sensing and control. The magnitude of forces involved in micromanipulation may range from 10−6 down to 10−9 N and below. Currently, the types of widely used microforce sensors are strain gauge, piezoelectric, capacitive and optical sensors. Strain gauges are easy to use, but their proper placement is crucial to obtaining good results. Piezoelectric force sensors are best at detecting forces very quickly, but are not fit for static force measurement. Capacitive force sensors are more sensitive than the two types discussed before, but their measurement range is usually limited. Finally, atomic force microscopy (AFM) has the highest sensitivity and may be used in noncontact force detection, but it has rather limited measurement range and requires additional force signal compensation [140–148].

Brunetto et al. [157] showed the characteristics of IPMC changes with the change in temperature and humidity. They developed a model of IPMC with applied voltage and current: One portion of the current produces charge redistribution and a stress inside the IPMC. This stress can be distributed into two components: a blocked force and a free deflection [21,118,158–168].

7.8.3.2 Comparison between Active Sensing, Passive Sensing and Self-Sensing Actuation (SSA)

Depending on the different IPMC sensing phenomena, sensing methods are divided into active sensing, passive sensing and self-sensing actuation (SSA). The active sensing methods measure one of the following: IPMC-generated voltage, charge or current; the passive methods measure variations in IPMC impedances, or are used in capacitive sensor element circuits; and the SSA methods implement simultaneous sensing and actuation on the same IPMC sample. The active IPMC sensing methods rely on IPMC mechanoelectrical transduction, i.e. the generation of electrical energy in response to bending. The passive IPMC sensing methods obtain information about material mechanical deformations by passing externally generated signals through (parts of) the IPMC sample and measuring alterations in this signal due to variations in the material electrical properties upon bending. The SSA methods for IPMC sensing aim for realizing sensing functionality within the actuating material sample. The reported active methods for sensing IPMC deformations can be separated into three categories: (1) measuring the voltage between IPMC electrodes (i.e. at high input impedance); (2) measuring the current between IPMC electrodes in virtual short-circuit configuration (i.e. at near-zero loading impedance); and (3) measuring the charge that is displaced between IPMC electrodes in virtual short-circuit configuration. The passive sensing methods rely on variations in IPMC-related electrical impedance properties in response

Polar Atmospheric Electric Field & IPMCs

to material deformations and require a measurement circuit that also supplies the power needed to detect these changes. The reported methods measure either variations in capacitances between IPMC electrodes and external metal plates or variations in the impedance of IPMC electrodes. The SSA methods achieve sensing and actuation simultaneously within the same material sample. The advantages of SSA methods include elimination of additional sensor weight, low volume and integration with cost-effectiveness. By working principle, the reported IPMC self-sensing actuation methods can be distinguished as: (1) methods that measure actuator electrode impedance; (2) methods that measure impedance through IPMC actuator; (3) methods that measure charge on IPMC actuator; and (4) methods that introduce separate actuator and sensor regions on the same IPMC sample [169–174].

7.8.3.3 Performance of IPMC Sensors

Chen et al. [175] showed that the polydimethylsiloxane (PDMS) and IPMC can be used together for actuation. The combination of IPMC actuator and PDMS membrane is used in the swimming robot 'manta ray' [109]. The IPMC membrane actuator size is 70 mm×80 mm. A total of 16 IPMCs are used for this design, and the total system is 18 cm×8 cm. The control result shows that the system is stable as the Bode plot of this system is stable. In the magnitude plot, the gain is below 0 dB throughout the frequency range and the gain margin and phase margin are positive. The response of the tip force measurement shows that it is a first-order system as it behaves as a low-pass filter. So, it is shown that the IPMC–PVDF membrane system is capable of 3D kinematic motion and consumes low power, i.e. less than 1 W. Min et al. [176] described the fabrication process of IPMCs. Also, they had improved the blocking force and displacement of IPMCs from 14.80 mN and 13.96 mm to 49.98 mN and 14.10 mm, respectively. They showed three aspects (metal electrode, hydration process and electromechanical performance) to explain how the blocking force of IPMCs can be increased. The EAPs can be used to develop artificial muscles with the potential of developing biologically inspired robots that can possibly walk, fly, hop, dig, swim and/or dive [111–116]. Different types of EAPs exist: wet (ionic) EAP ionic polymers, IPMCs, carbon nanotubes, dry EAP ferroelectric polymers (PVDF or PVF2), electrostatically stricted polymer (ESSP) actuators and electro-viscoelastic elastomers [149]. Lee et al. [177] showed the performance improvement of IPMCs on the solvent loss and actuation force [124]. For the reduction in solvent loss and improvement in actuation force, IPMCs with various solvents (DI water, D2O and DMSO) and cations (H+, Li+ and Cu++) were tested for the measurement of solvent loss and actuation force. They tested the wing flapping device using an IPMC of 800 µm thickness, and the rack and pinion could create around 10°~85° flapping angles and 0.5~15 Hz flapping frequencies by applying 3~4 V. Jung et al. [178] showed the performance of the IPMC actuator according to varying frequencies of actuation voltages, where the size of actuator was 10 mm (length)×2 mm (width)×0.2 mm (thickness) and a square input voltage of ±1.5 V_{pp} was applied. The IPMC actuator generates most of its deformation just after the polarity of the excitation voltage is switched and it becomes the most of its overall deformation. Different signals such as square, sinusoidal and triangular waves may also be applied. The square input consumes quite a larger amount of power than the others. Driving the IPMC actuator by the sinusoidal input is more efficient than using a square one with respect to power consumption [130–132]. The forces are not much different among the waveforms when a driving voltage ±3 V_{pp} is applied. Feng et al. [179] gave a new idea of fabricating IPMC and designed a μIPMC for biological applications. The dimensions of the fabricated μIPMC were 6 mm × 6 mm × 80 μm. The displacement and force outputs were 300 μm and 5 mN, respectively. The supply given to the system was a 12 V, 0.5 Hz square wave. This μIPMC gripper grips a flexible tube of 800 μm. Tsiakmakis et al. [180] developed a motion parameter with a charge-coupled device (CCD) video camera. First capture the motion in the video, then transfer the captured video in frames, and then analyse the frames. The camera-based measurement gave a more prominent result than the laser displacement measurement. Applying a frequency of 0–7 Hz, the maximum displacement observed was 0–3.5 mm. This method is quite suitable for underwater actuation measurement. Sadeghipou et al. [181] reported a smart material sensor named Nafion. Assuming the ideal gas behaviour of hydrogen, the cell produces an open-circuit voltage given by the Nernst equation as

$$E = \frac{RT}{2F} \ln \frac{P_1}{P_2} \qquad (7.3)$$

where R = the gas constant (8.314 J/K), F = the Faraday constant (96,500 C), T = the absolute temperature (K), and P_1, P_2 = the pressures on the two sides of the Nafion. Lee et al. [182] performed a mechanical structural analysis of IPMC using the finite element analysis programme ANSYS. A comparative study of the experimental result with IPMC displacement and force with the ANSYS result was done in the report. For the experiment, they took IPMCs of different thicknesses, and it can be observed from the table that the force increases and the displacement decreases with the increase in IPMC thickness. Chen et al. [183,184] reported on the Pt–Ag and Ag–Ag electroding on IPMC. The performance in this electroding showed an improved actuation of IPMC, i.e., Pt–Ag electrode IPMC shows better actuation than Pt electrode IPMC. They also presented the preparation process of Pt–Ag and Ag–Ag IPMCs. The Ag–Ag IPMC had the smallest and most stable surface resistance than the Pt-plated one. Chen et al. [185] used the IPMC as a flow sensor. They reported that Young's modulus of the IPMC was 500 N/mm^2 and the density of the IPMC strip of dimensions 2×4×0.25 mm was 3 g/cm^3 [21,107,117,186–198].

7.9 THE PROPOSED IPMC-BASED METHODOLOGY (WEARABLE ELECTRONIC DEVICE PROTOTYPE DESIGN)

Gonzalez et al. [199] analysed the present and future prospects of energy harvesting from mechanical actuation in the human gesture using the piezoelectric effect. They also discussed the portable wearable electronics devices in view of energy autonomous systems. They also discussed the feasibility of the energy harvested from human body to power wearable autonomous systems. Shoe-mounted rotary electromechanical generators produce power from walking to supply these devices. The power level obtained in this study is based on theoretical calculations. The basic features of a wearable device include programmability with microprocessors, wireless connectivity such as Bluetooth and advanced GSM terminal with extended functions. The general power and energy trends of wearable devices are component scaling down trend following the Moore's law and power consumption scaling down trend.

Meddad et al. [200] presented the power analytical model developed by a smart structure of PZT (piezoelectric ceramic material) which is used as supply energy for electronic device. Two vibration modes (transverse and longitudinal) were used and compared. To maximize the electric power, the longitudinal mode requires a small surface, a great thickness and a maximum number of layers. The limitation of this mode is that the mechanical force applied to a small surface to generate the EMF is difficult to integrate in MEMS. In transverse mode, the number of layers is maximized to maximize the generated electric power [201–205].

7.10 DISCUSSION AND CONCLUSIONS

In this chapter, we have briefly summarized the electrical nature of the earth's atmosphere and natural sources of electromagnetic energy near the earth's surface and in the ionosphere/magnetosphere. The transport of energy from one region to the other region is also discussed. The atmospheric electric field and its diurnal variation with respect to definite environmental aspects are elaborately discussed. Further, the naturally occurring potential gradient in the Polar Regions and the impacts, i.e. several physiological effects on the people residing there, are summarized in this review work. In fact, further research is necessary to better understand the natural environment and thereby the existing atmospheric EFs in the Polar Regions and its impact on human life so that future technological evolution may be established. The authors endorse that the EAP like IPMC can be the solution for the discharging of resounding charge carrying out by the human being in the Polar Region like Antarctica. This review chapter intends to act as an extensive comprehensive reading material for research scholars and developers working in this field (Table 7.3).

TABLE 7.3
Summary of Research Articles Reported in the Literature Review

Topic	Ref. No.	Author	Year	Research Findings
Atmospheric electric field in Polar Regions	[1]	Dolezalek	June 1971	Reported the atmospheric electricity and showed the evidence of global electricity and global electric circuit.
	[2]	Edgar A. Bering et al.	1998	Described the global electric circuit where the source of ionization is the collision between the galactic cosmic rays and the neutral molecules present in the atmosphere.
	[3]	Graham J. Hillused	2020	On the use of electromagnetics for earth imaging of the Polar Regions.
	[4]	Devendraa Siingh et al.	2013	Explored various atmospheric electric parameters and the diurnal variation of fair-weather electric fields at Maitri station (the Indian Antarctic station on the Antarctic plateau, 130 m above mean sea level).
	[6]	Kamra et al.	1994	Investigated the unitary diurnal variation of the atmospheric electric field; their observation was based on the regions of Indian Ocean, Bay of Bengal and Arabian Sea.
	[7]	Guha et al.	2010	Measured the fair-weather atmospheric electricity in the Northeast India.
	[8]	Deshpande and Kamra	2001	Reported the diurnal variation of atmospheric electricity and the potential gradient during 20 fair-weather days at the Indian Station, Maitri, located at Antarctica.
	[9]	Jeni Victor et al.	2015	Discussed the temporal variation of surface electric field measured at Maitri and Vostok and its departure during the geomagnetic-disturbed periods.
	[10]	Masashi Kamogawa	2015	Monitoring atmospheric electricity in mountain areas at Mount Fuji, Japan.

(Continued)

TABLE 7.3 (*Continued*)
Summary of Research Articles Reported in the Literature Review

Topic	Ref. No.	Author	Year	Research Findings
	[11]	Masashi Kamogawa et al.	2015	Diurnal variation of atmospheric electric field at the summit of Mount Fuji, Japan, distinctly different from the Carnegie curve in the summertime.
	[23]	M.J. Rycroft et al.	2000	Investigated several factors associated which influences that such as lightning and thunderstorms, sprites, elves and blue jets, those are having direct effect on the global circuit.
	[24]	Damian Murphy et al.	2020	Described the presence of earth's vertical electric field in the high altitude of Antarctic plateau.
	[25]	Aplin et al.	2013	Described different measuring devices and the methodology applied first by Lord Kelvin on atmospheric electricity.
	[26]	Harrison	2004	Described atmospheric electricity and the global electric circuit and the effects of several atmospheric parameters.
	[27]	Hutchinson	1951	They found an approximate empirical relationship: $i = a$ (FZ−MZ), where i is the downward directed current in pA, F the downward directed field in v/cm, M the minimum field for the onset of point discharge, and 'a' the constant depending on the sign of current.
	[29]	De et al.	2013	Studied the atmospheric electric potential gradient at Kolkata during solar eclipse on 22 July 2009.
	[30]	Adarsh Kumar et al.	2013	Described the impact of high-energy cosmic rays on global atmospheric electrical parameters over different orographically important places of India.
	[31]	Earle R. Williams	2009	Reviewed the global electrical circuit.

(*Continued*)

TABLE 7.3 (*Continued*)
Summary of Research Articles Reported in the Literature Review

Topic	Ref. No.	Author	Year	Research Findings
Air–earth current	[28]	Valeo et al.	2019	Done a preliminary study of fair-weather atmospheric electric field at a high-altitude station (Shillong) – 25°N, 91°E.
	[35]	Singh et al.	2004	Reviewed the electrical environments and interconnection through electrodynamic coupling in view of Maxwell current.
	[40]	Lothar H. Ruhnke	1969	Described the area averaging of atmospheric electric currents.
Diurnal variation of atmospheric parameters	[41]	Richard J. Blakeslee et al.	2014	Established that the electrified storms are the source of the fair-weather electric field.
	[42]	Latha	2003	Explained the diurnal variation of surface electric field at a tropical station in different seasons.
	[43]	Yifan Wang et al.	2018	Investigated the temporal variation of atmospheric static electric field due to air ions and their relationships to pollution in Shanghai.
	[46]	Burns et al.	2005	Demonstrated the inter-annual consistency of average diurnal vertical electric field.
	[47]	Harrison	2011	Studied the origin of the fair-weather atmospheric electric field which was first introduced by Lord Kelvin using electrostatic instrumentation.
	[48]	Bennett et al.	2007	Observed the atmospheric electrical parameters at different weather conditions such as fair-weather conditions and disturbed weather.

(*Continued*)

TABLE 7.3 (Continued)
Summary of Research Articles Reported in the Literature Review

Topic	Ref. No.	Author	Year	Research Findings
Static electric field and physiological effects on human beings	[64]	Jonathan M. Charry	1984	Biological effects of small air ions.
	[65]	Knoll et al.	1984	Properties, measurement and bioclimatic action of "small" multimolecular atmospheric ions.
	[66]	Breton et al.	1998	Atmospheric ionization patterns at 4 m above ground level in relation to meteorological events.
	[67]	Bachman et al.	1965	Physiological effects of measured air ions.
	[68]	Dawson et al.	2004	Interactions of electric fields due to electrostatic discharge with human tissue.
	[69]	Kai Wang et al.	2003	Numerical modelling of electrostatic discharge generators.
	[70]	William T. Kaune	1981	Detection of power-frequency electric field effect over the body surfaces of grounded humans and animals.
Types of EAPs and selection criteria as the solution of body static nullification technology	[15]	Bar-Cohen	2017	Electroactive polymers as actuators.
	[16]	Kwang J. Kim et al.	2002	Ionic polymer–metal composites: manufacturing techniques.
	[17]	Min Yu et al.	2007	Manufacture and performance of ionic polymer–metal composites.
	[18]	Yoseph Bar-Cohen et al.	2008	Electroactive polymer actuators and sensors.
	[101]	Qingguo Li et al.	2009	Development of a biomechanical energy harvester.
	[102]	Donelan et al.	2008	Biomechanical energy harvesting.
	[103]	Farrell et al.	2008	Concerning the dissipation of electrically charged objects in the shadowed lunar Polar Regions.
	[104]	Pavlos K. Katsivelis	2010	Electrostatic discharge current linear approach and circuit design method.
	[105]	Mohsen Shahinpoor et al.	2001	Ionic polymer–metal composites and their properties.

(Continued)

TABLE 7.3 (Continued)
Summary of Research Articles Reported in the Literature Review

Topic	Ref. No.	Author	Year	Research Findings
IPMC-based technologies – the proposed methodology	[203]	Shahinpoor et al.	1998	Ionic polymer–metal composites (IPMCs) as biomimetic sensors, actuators and artificial muscles.
	[205]	Byoung Hun Kang et al.	2006	Design of compliant MEMS grippers for microassembly tasks.
	[206]	Liviu-Mircea Enache	2019	Air ionization – an environmental factor with therapeutic potential.
	[207]	Albrechtsen et al.	1978	The influence of small atmospheric ions on human well-being and mental performance.
	[208]	Brunetto et al.	2009	Static and dynamic characterization of the temperature and humidity influence on IPMC actuators.
	[209]	Brufau-Penella et al.	2008	Characterization of the harvesting capabilities of an ionic polymer–metal composite device.
	[210]	Srijan Bhattacharya et al.	2014	IPMC-actuated compliant mechanism-based multifunctional, multifinger microgripper.

REFERENCES

[1] H. Dolezalek, Atmospheric electricity, 52, 6, pp. 351–368, June 1971, https://agupubs.onlinelibrary.wiley.com/doi/abs/10.1029/EO052i006pIU351.

[2] E.A. Bering, A.A. Few, J.R. Benbrook, The global electric circuit, American Institute of Physics, doi: 10.1063/1.882422, 51, pp. 24–30, 1998.

[3] G.J. Hill, On the use of electromagnetics for earth imaging of the polar regions, *Surveys in Geophysics*, 41, pp. 5–45, 2020, doi: 10.1007/s10712-019-09570-8.

[4] D. Siingh, R.P. Singh, V. Gopalakrishnan, C. Selvaraj, C. Panneerselvam, Fair-weather atmospheric electricity study at Maitri (Antarctica), *Earth, Planets and Space*, 65, pp. 1541–1553, December 2013, doi: 10.5047/eps.2013.09.011.

[5] R. Giles Harrison, G.J. Marlton, Fair weather electric field meter for atmospheric science platforms, *Journal of Electrostatics*, 107, 103489, 2020, doi: 10.1016/j.elstat.2020.103489.

[6] A.K. Kamra, C.G. Deshpande, V. Gopalakrishna, Challenge to the assumption of the unitary diurnal variation of the atmospheric electric field based on observations in the Indian Ocean, Bay of Bengal, and Arabian Sea, *Journal of Geophysical Research: Atmospheres*, 99, 10, pp. 21043–21050, October 1994. https://agupubs.onlinelibrary.wiley.com/doi/abs/10.1029/94JD01485, doi: 10.1029/94JD01485.

[7] A. Guha, B.K. De, S. Gurubaran, S.S. De, K. Jeeva, First results of fair-weather atmospheric electricity measurements in northeast India, *Journal of Earth System Science*, 119, pp. 221–228, January 2010, doi: 10.1007/s12040-010-0014-9.

[8] C.G. Deshpande, A.K. Kamra, Diurnal variations of the atmospheric electric field and conductivity at Maitri, Antarctica, *Journal of Geophysical Research*, 106, 13, pp. 14207–14218, July 2001, doi: 10.1029/2000JD900675.

[9] N.J. Victor, C. Panneerselvam, C.P.A. Kumar, Variation of surface electric field during geomagnetic disturbed period at Maitri, Antarctica, *Journal of Earth System Science*, 124, 8, pp. 1721–1733, December 2015, doi: 10.1007/s12040-015-0638-x.

[10] M. Kamogawa, *Monitoring Atmospheric Electricity in Mountain Areas*, Tokyo Gakugei University, Chapter 4.6.

[11] M. Kamogawa et al., Diurnal variation of atmospheric electric field at the summit of Mount Fuji, Japan, distinctly different from the Carnegie curve in the summertime, *Geophysical Research Letters*, 2, 8, pp. 3019–3023, April 2015, doi:10.1002/2015GL063677.

[12] Y.M. Choi, M. Gu Lee, Y Jeon, Wearable biomechanical energy harvesting technologies, *Energies*, 10, 1483, 2017, doi: 10.3390/en10101483.

[13] N. Jonassen, Neutralization of static charges by air ions: part I, theory, *Mr. Static, Compliance Engineering*, 19, 2, pp. 28–31, 2002.

[14] E. Okoniewska, M.A. Stuchly, M. Okoniewski, Interactions of electrostatic discharge with the human body, *IEEE Transactions on Microwave Theory and Techniques*, 52, 8, pp. 2030–2039, August 2004.

[15] Y. Bar-Cohen, V.F. Cardoso, C. Ribeiro, S. Lanceros-Méndez, Electroactive polymers as actuators, *ScienceDirect, Advanced Piezoelectric Materials (Second Edition), Science and Technology*, Woodhead Publishing in Materials 2017, Book Chapter 8–Electroactive Polymers as Actuators, doi: 10.1016/B978-0-08-102135-4.00008-4, pp. 319–352, 2017.

[16] K.J. Kiml, M. Shahinpoor, Ionic polymer-metal composites: manufacturing techniques, SPIE's 9th Annual International Symposium on Smart Structures and Materials, 2002, San Diego, California, United States, Smart Structures and Materials 2002: Electroactive Polymer Actuators and Devices (EAPAD), doi: 10.1117/12.475166, 4695, pp. 210–219, July 2002.

[17] M. Yu, H. Shen, Z. Dai, Manufacture and performance of ionic polymer-metal composites, *Journal of Bionic Engineering*, 4, 3, pp. 143–149, September 2007, doi: 10.1016/S1672-6529(07)60026-2.
[18] Y.B. Cohen, Q. Zhang, Electroactive polymer actuators and sensors, *MRS Bulletin*, 33, 3, pp. 173–181, March 2008, doi: 10.1557/mrs2008.42.
[19] Y.B. Cohen, Biomimetics: biologically inspired technology, II ECCOMAS Thematic Conference on Smart Structures and Materials Lisbon, Portugal, C.A. Mota Soares et al. (Eds.). https://www.researchgate.net/publication/242526086, July 2005.
[20] A. Hunt, A. Punning, M. Anton, A. Aabloo, M. Kruusmaa, A multilink manipulator with IPMC joints, Electroactive Polymer Actuators and Devices (EAPAD) 2008, Proc. of SPIE, doi: 10.1117/12.775952, 692769271-1, 2008.
[21] Y.B. Cohen, S. Sherrit, S.S. Lih, Characterization of the electromechanical properties of EAP materials, Smart Structures and Materials 2001: Electroactive Polymer Actuators and Devices, Proceedings of SPIE, Proc. SPIE 4329, pp. 319–327, 2001.
[22] M. Shahinpoor, Y. Bar-Cohen, T. Xue, J.O. Simpson, J. Smith, Ionic polymer-metal composites (IPMC) as biomimetic sensors and actuators, Proceedings of SPIE's 5th Annual International Symposium on Smart Structures and Materials, 1–5 March, 1998, San Diego, CA. pp. 3324–27, 1998.
[23] M.J. Rycroft, S. Israelsson, C. Price, The global atmospheric electric circuit, solar activity and climate change, *Journal of Atmospheric and Solar-Terrestrial Physics*, 62, 17–18, pp. 1563–1576, November 2000, doi: 10.1016/S1364-6826(00)00112-7.
[24] D. Murphy, J. Denny, Earth's Vertical Electric Field.
[25] K.L. Aplin, R.G. Harrison, Lord Kelvin's atmospheric electricity measurements, *History of Geo- and Space Sciences*, 4, pp. 83–95, September 2013, www.hist-geo-space-sci.net/4/83/2013/, doi: 10.5194/hgss-4-83-2013.
[26] R.G. Harrison, The global atmospheric electrical circuit and climate, *Surveys in Geophysics*, 25, pp. 441–484, November 2004, doi: 10.1007/s10712-004-5439-8.
[27] W.C.A. Hutchinson, Point-discharge currents and the earth's electric field, *Quarterly Journal of the royal Metrological Society*, 77, 334, pp. 627–632, October 1951, doi: 10.1002/qj.49707733406.
[28] V. Valeo, P.V. Koparkar, A preliminary study of fair-weather atmospheric electric field at a high altitude station (Shillong) 25°N, 91°E, *International Journal of Applied Engineering Research*, 14, 7, pp. 1652–1657, 2019, ISSN 0973-4562.
[29] S.S. De, B. Bandyopadhyay, S. Paul, D.K. Haldar, M. Bose, D. Kala, G. Guha, Atmospheric electric potential gradient at Kolkata during solar eclipse of 22 July 2009, *Indian Journal of Radio & Space Physics*, 42, pp. 251–258, August 2013.
[30] A. Kumar, H.P. Singh, Impact of high energy cosmic rays on global atmospheric electrical parameters over different orographically important places of India, *Hindawi Publishing Corporation ISRN High Energy Physics*, 2013, Article ID 831431, 7 pages, doi: 10.1155/2013/831431.
[31] E.R. Williams, The global electrical circuit: a review, *Atmospheric Research*, 91, pp. 140–152, 2009, doi: 10.1016/j.atmosres.2008.05.018.
[32] S.V. Anisimov, E.A. Mareev, S.S. Bakastov, On the generation and evolution of aero-electric structures in the surface layer, *Journal of Geophysical Research*, 104, 12, pp. 14,359–14,367, June 27, 1999.
[33] M.M. Lam et al., Solar wind-atmospheric electricity-cloud microphysics connections to weather and climate, *Journal of Atmospheric and Solar-Terrestrial Physics*, 149, pp. 277–290, 2019.
[34] E.J. Adlerman, E.R. Williams, Seasonal variation of the global electrical circuit, *Journal of Geophysical Research*, 101, 23, pp. 679–688, December 1996, doi: 10.1029/96JD01547.

[35] D.K. Singh, R.P. Singh, A.K. Kamra, The electrical environment of the earth's atmosphere: a review, *Space Science Reviews*, 113, pp. 375–408, August 2004, https://link.springer.com/article/10.1007/s11214-005-0853-x.
[36] N. Jonassen, Environmental ESD: Part 1 – The Atmospheric Electric Circuit, 16, 3, pp. 24–28, 1999.
[37] T. Ogawa, *Analyses of Measurement Techniques of Electric Fields and Currents in the Atmosphere*, Geophysical Institute, Kyoto University, 13, pp. 111–137, 1973.
[38] R.G. Harrison, Fair weather atmospheric electricity: its origin and applications, Proc. ESA Annual Meeting on Electrostatics, 2014.
[39] F.J. Welsh Whipple, F.J. Scrase, Point discharge in the electric field of the earth, an analysis of continuous records obtained at Kew observatory, H. M. Stationery Office, 1936–Atmospheric Electricity, https://books.google.co.in/books/about/PointDischargein_the_Electric_Field_of.html, 1936.
[40] L.H. Ruhnke, Area averaging of atmospheric electric currents, *Journal of Geomagnetism and Geoelectricity*, 21, 1, 1969.
[41] R.J. Blakeslee, D.M. Mach, M.G. Bateman, J.C. Bailey, Seasonal variations in the lightning diurnal cycle and implications for the global electric circuit, *Atmospheric Research*, 135–136, pp. 228–243, January 2014, doi: 10.1016/j.atmosres.2012.09.023.
[42] R. Latha, Diurnal variation of surface electric field at a tropical station in different seasons: a study of plausible influences, *Earth Planets Space*, 55, pp. 677–685, 2003.
[43] Y. Wang et al., Temporal variation of atmospheric static electric field and air ions and their relationships to pollution in Shanghai, *Aerosol and Air Quality Research*, 18. pp. 1631–1641, 2018, doi: 10.4209/aaqr.2017.07.0248.
[44] E.R. Williams, S.J. Heckman, The local diurnal variation of cloud electrification and the global diurnal variation of negative charge on the earth, *Journal of Geophysical Research*, 98, 3, pp. 5221–5234, March 20, 1993.
[45] T.D. Bracken, A.S. Capon, D.V. Montgomery, Ground level electric fields and ion currents on the Celilo-Sylmar ±400 kv dc intertie during fair weather, *IEEE Transactions on Power Apparatus and Systems*, 97, 2, March/April 1978.
[46] G.B. Burns, A.V. Frank-Kamenetsky, O.A. Troshichev, E.A. Bering, B.D. Reddell, Inter-annual consistency of seasonal differences in diurnal variations of the geoelectric field, *Journal of Geophysical Research Atmospheres*, 110, D10106, May 2005, doi: 10.1029/2004JD005469.
[47] R.G. Harrison, Fair weather atmospheric electricity, 13th International Conference on Electrostatics, Journal of Physics: Conference Series 301, 012001, 2011, doi:10.1088/1742-6596/301/1/012001.
[48] A.J. Bennett, R.G. Harrison, Atmospheric electricity in different weather conditions, *Weather*, 62, 10, pp. 277–283, October 2007, doi: 10.1002/wea.97.
[49] J.C. Willett, Bailey, J.C., Contact-Potential and Surface-Charge Effects in Atmospheric-Electrical Instrumentation, 1983STIN...8332027W, April 1983.
[50] G.J. Byrne, J.R. Benbrook, E.A. Bering, A.A. Few, G.A. Morris, W.J. Trabucco, E.W. Paschal, Ground-based instrumentation for measurements of atmospheric conduction current and electric field at the south pole, *Journal of Geophysical Research*, 98, 2, pp. 2611–2618, February 1993, doi: 10.1029/92JD02303.
[51] Y. Morita, H. Ishikaw, Simultaneous measurements of electric conductivity and aerosol in the lower stratosphere, *Journal of Geomagnetism and Geoelectricity*, 28, pp. 09–315, 1976.
[52] A.K. Kamra, Measurements of the electrical properties of dust storms, *Journal of Geophysical Research*, 77, 30, pp. 5856–5869, October 1972, doi: 10.1029/JC077i030p05856.

[53] D. Lubin, J.E. Frederick, The ultraviolet radiation environment of the Antarctic Peninsula: the roles of ozone and cloud cover, *Journal of Applied Meteorology*, 30, 4, pp. 478–493, April 1991, doi: 10.1175/1520-0450(1991)030<0478:TUREOT>2.0.CO; 2.
[54] D. Siingha, R.P. Singhb, A.K. Kamraa, P.N. Guptab, R. Singhb, V. Gopalakrishnana, A.K. Singh, Review of electromagnetic coupling between the earth's atmosphere and the space environment, *Journal of Atmospheric and Solar-Terrestrial Physics*, 67, pp. 637–658, September 2004, doi: 10.1016/j.jastp.2004.09.006.
[55] V.V. Kumar, V. Ramachandran, V. Buadromo, J. Prakash, Surface fair-weather potential gradient measurements from a small tropical island station Suva, Fiji, *Earth Planets and Space*, 61, 6, pp. 747–753, July 2009, doi: 10.1186/BF03353181.
[56] J.M. Rosen et al., Results of an international workshop on atmospheric electrical measurements, *Journal of Geophysical Research*, 87, 2, pp. 1219–1227, February 1982.
[57] K.A. Nicoll, Measurements of atmospheric electricity aloft, *Surveys in Geophysics*, 33, pp. 991–1057, 2012, doi: 10.1007/s10712-012-9188-9.
[58] C.G Deshpande, AK Kamra, The atmospheric electric field and conductivity measurements during the XVI Indian Antarctica Expedition, Sixteenth Indian Expedition to Antarctica, Scientific Report, Department of Ocean Development, Technical Publication No. 14, pp. 138–152, 2000.
[59] G.B. Burns, M.H. Hesse, S.K. Parcell, S. Malachowskit, K.D. Colef, The geoelectric field at Davis station, Antarctica, *Journal of Atmospheric and Terrestrial Physics*, 57, 14, pp. 1783–1797, December 1995, doi: 10.1016/0021-9169(95)00098-M.
[60] Wikipedia, Atmospheric Electricity, pp. 1–11, 2021, https://en.wikipedia.org/wiki/Atmospheric_electricity, Wikimedia Commons.
[61] R. Markson, R. Nelson, Mountain-peak potential-gradient measurements and the Andes glow, *Weather*, 25, 8, pp. 350–360, August 1970, doi: 10.1002/j.1477-8696.1970.tb04118.x.
[62] C.B. Boyle, P.H. Reiff, M.R. Hairston, Empirical polar cap potentials, *Journal of Geophysical Research*, 102, 1, pp. 111–125, January 1997, doi: 10.1029/96JA01742.
[63] Wikipedia, Static Electricity pp. 1–11, 2021, https://en.wikipedia.org/wiki/Static_electricity,Wikimedia Commons.
[64] J.M. Charry, Biological effects of small air ions: a review of findings and methods, *Environmental Research*, 34, pp. 351–389, 1984.
[65] M. Knoll, J. Eichmeier, R.W. Schun, *Properties, Measurement, and Bioclimatic Action of "Small" Multimolecular Atmospheric Ions*, Institut fur Technische Elektronik, Technische Hochschule Munchen, Munchen, Germany, 6, pp. 177–254.
[66] J. Breton, V. Breton, Y. Legoff, Atmospheric ionization patterns at 4 m above ground level in correlation to meteorological events, *Journal of Geophysical Research*, 103, 2, pp. 1837–1846, January 1998.
[67] C.H. Bachman, R.D. McDonald, P.J. Lorenz, Some physiological effects of measured air ions, *International Journal of Biometeorology*, 9, 2, pp. 127–139, 1965.
[68] T.W. Dawson, M.A. Stuchly, R. Kavet, Evaluation of interactions of electric fields due to electrostatic discharge with human tissue, *IEEE Transactions on Biomedical Engineering*, 51, 12, December 2004.
[69] K. Wang, D. Pommerenke, R. Chundru, T.V. Doren, J.L. Drewniak, A. Shashindranath, Numerical modeling of electrostatic discharge generators, *IEEE Transactions On Electromagnetic Compatibility*, 45, 2, pp. 258–271, May 2003.
[70] W.T. Kaune, Power-frequency electric fields averaged over the body surfaces of grounded humans and animals, *Bio-Electromagnetics*, 2, pp. 403–406, 1981.

[71] W.H. Bailey, J.M. Charry, Behavioral monitoring of rats during exposure to air ions and dc electric fields, *Bio-Electromagnetics*, 7, pp. 329–339, 1986.
[72] M. Ghaly, D. Teplitz, The biologic effects of grounding the human body during sleep as measured by cortisol levels and subjective reporting of sleep, pain, and stress, *The Journal of Alternative and Complementary Medicine*, 10, 5, pp. 767–776, 2004.
[73] R. Applewhite, The effectiveness of a conductive patch and a conductive bed pad in reducing induced human body voltage via the application of earth ground, *European Biology and Bioelectromagnetics*, 1, pp. 23–40, 2005.
[74] I.A. Jamieson, S.S. Jamieson, H.M. ApSimon, J.N.B. Bell, Grounding & human health – a review, Journal of Physics: Conference Series 301, 2011, 012024, doi: 10.1088/1742-6596/301/1/012024.
[75] A.K. Petri, K. Schmiedchen, D. Stunder, D. Dechent, T. Kraus, W.H. Bailey, S. Driessen, Biological effects of exposure to static electric fields in humans and vertebrates: a systematic review, *Environmental Health*, 16, Article number 41, 2017, doi: 10.1186/s12940-017-0248-y.
[76] J.P. Blondin, D.H. Nguyen, J. Sbeghen, D. Goulet, C. Cardinal, P.S. Maruvada, M. Plante, W.H. Bailey, Human perception of electric fields and ion currents associated with high- voltage DC transmission lines, *Bioelectromagnetics*, 17, pp. 230–241, 1996.
[77] H.O. Shimizu, K. Shimizu, Experimental analysis of the human perception threshold of a dc electric field, *Medical & Biological Engineering & Computing*, 37, pp. 727–732, 1999.
[78] T.W. Dawson, M.A. Stuchly, R. Kavet, Electric fields in the human body due to electrostatic discharges, *IEEE Transactions on Biomedical Engineering*, 51, 8, pp. 1460–1468, August 2004.
[79] W. Li, Q. Li, J. Li, Influence of human body static elimination technology on human electrostatic potential, 2017 8th International Conference on Fire Science and Fire Protection Engineering (on the Development of Performance-based Fire Code), Procedia Engineering, 211, doi: 10.1016/j.proeng.2017.12.040, pp. 488–494, 2018.
[80] J. Bowman, J. Niple, R. Kavet, Pilot measurements of elf contact currents in some electric utility occupations, *Journal of Occupational and Environmental Hygiene*, 3, pp. 323–333, June 2006, doi: 10.1080/15459620600697642.
[81] A.P. Krueger, Are air ions biologically significant? A review of a controversial subject, *International Journal of Biometeorology*, 16, 4, pp. 313–322, 1972.
[82] G. Chevalier, S.T. Sinatra, J.L. Oschman, K. Sokal, P. Sokal, Earthing: health implications of reconnecting the human body to the earth's surface electrons, *Journal of Environmental and Public Health*, 2012, Article ID 291541, 8 pages, October 2011, doi: 10.1155/2012/291541.
[83] T.W. Dawson, K. Caputa, M.A. Stuchly, R. Kavet, Electric fields in the human body resulting from 60-hz contact currents, *IEEE Transactions on Biomedical Engineering*, 48, 9, pp. 1020–1026, September 2001.
[84] L. Xinggang, W. Ming, H. Xiaofeng, Mathematical method for air electrostatic discharge circuits calculation, *High Voltage*, 3, 3, pp. 226–231, June 2018, doi: 10.1049/hve.2017.0157.
[85] S. Gabriely, R.W. Lau, C. Gabriel, The dielectric properties of biological tissues: II. Measurements in the frequency range 10 Hz to 20 GHz, *Physics in Medicine and Biology*, 41, Printed in the UK, pp. 2251–2269, April 1996.
[86] M.H. Repacholi, B. Greenebaum, Interaction of static and extremely low frequency electric and magnetic fields with living systems: health effects and research needs, *Bioelectromagnetics*, 20, pp. 133–160, June 1998.

[87] G. Chevalier, S.T. Sinatra, J.L. Oschman, K. Sokal, P. Sokal, Earthing: health implications of reconnecting the human body to the earth's surface electrons, *Journal of Environmental and Public Health*, 2012, Article ID 291541, 8 pages, doi: 10.1155/2012/291541.

[88] K.K. Katrak, Human body electrostatic charge (ESC) levels: are they limited by corona bleed off or environmental conditions, EOS/ESD Symposium 95-73 to 95-85, pp. 2.3.1–2.3.13.

[89] P.E. Sutherland, D. Dorr, K. Gomatom, Human current sensitivities and resistance values in the presence of electrically energized objects, I&CPS -05–19, pp. 150–158.

[90] K. Chamberlin, W. Smith, C. Chirgwin, S. Appasani, P. Rioux, Analysis of the charge exchange between the human body and ground: evaluation of earthing from an electrical perspective, *Journal of Chiropractic Medicine*, 13, pp. 239–246, September 2014.

[91] N. Ichikawa, Investigation of human body potential measured by a non-contact measuring system, *Industrial Health*, 54, pp. 542–549, June 2016.

[92] C.I. Calle, The electrostatic environment at the international space station, Proc. 2016 Electrostatics Joint Conference, Purdue University, June 2016.

[93] R.K. Pandey, C.K. Ao, W. Lim, Y. Sun, X. Di, H. Nakanishi, S. Soh, The relationship between static charge and shape, *ACS Central Science*, 2020, 6, pp. 704–714, doi: 10.1021/acscentsci.9b01108.

[94] Electrostatic Discharge Capability of Clothing – Part One of Two, https://www.uvex-safety.com/blog/electrostatic-discharge-capability-of-clothing-part-one-of-two/, February 2017.

[95] Electrostatic Discharge Capability of Clothing – Part Two of Two, https://www.uvex-safety.com/blog/electrostatic-discharge-capability-of-clothing-part-two-of-two/, February 2017.

[96] J. Bernhardt, The direct influence of electromagnetic fields on nerve- and muscle cells of man within the frequency range of 1 Hz to 30 MHz, *Radiation and Environmental Biophysics*, 16, pp. 309–323, 1979.

[97] W.D. Schindler, P.J. Hauser, Antistatic finishes, Chemical Finishing of Textiles, https://www.sciencedirect.com/topics/engineering/antistatic-finish, 2004.

[98] R. Riemer, A. Shapiro, Biomechanical energy harvesting from human motion: theory, state of the art, design guidelines, and future directions, *Journal of Neuroengineering and Rehabilitation*, 8, 22, pp. 1–13, 2011.

[99] D.W. Kim, J.H. Lee, J.K. Kim, U. Jeong, Material aspects of triboelectric energy generation and sensors, *NPG Asia Materials*, 1, 2, 6, 2020, doi: 10.1038/s41427-019-0176-0.

[100] C. Bajrang, G.V. Suganthi, R. Tamilselvi, M. Parisabeham, A. Nagaraj, A systematic review of energy harvesting from biomechanical factors, *Biomedical & Pharmacology Journal*, 12, 4, pp. 2063–2070, December 2019.

[101] Q. Li, V. Naing, J.M. Donelan, Development of a biomechanical energy harvester, *Journal of Neuroengineering and Rehabilitation*, 6, pp. 1–12, June 2009, doi: 10.1186/1743-0003-6-22.

[102] J.M. Donelan, Q. Li, V. Naing, J.A. Hoffer, D.J. Weber, A.D. Kuo, Biomechanical energy harvesting: generating electricity during walking with minimal user effort, 319, pp. 807–810, February 2008.

[103] W.M. Farrell, T.J. Stubbs, G.T. Delory, R.R. Vondrak, M.R. Collier, J.S. Halekas, R.P. Lin, Concerning the dissipation of electrically charged objects in the shadowed lunar polar regions, *Geophysical Research Letters*, 35, pp. 1–5, 2008, doi: 10.1029/2008GL034785.

[104] P.K. Katsivelis, G.P. Fotis, I.F. Gonos, T.G. Koussiouris, I.A. Stathopulos, Electrostatic discharge current linear approach and circuit design method, *Energies*, 3, pp. 1728–1740, November 2010, doi: 10.3390/en3111728.
[105] M. Shahinpoor, K.J. Kim, Ionic polymer–metal composites: I. Fundamentals, *Smart Materials and Structures*, 10, 4, pp. 819–833, August 2001, doi: 10.1088/0964-1726/10/4/327.
[106] M. Shahinpoor, K.J. Kim, Ionic polymer–metal composites: iii. modeling and simulation as biomimetic sensors, actuators, transducers, and artificial muscles, *Smart Materials and Structures*, 13, 6, pp. 1362–1388, October 2004, doi: 10.1088/0964-1726/13/6/009.
[107] M. Shahinpoor, K.J. Kim, Ionic polymer–metal composites: iv. industrial and medical applications, *Smart Materials and Structures*, 14, 1, pp. 197–214, December 2004, doi: 10.1088/0964-1726/14/1/020.
[108] E. Malone, H. Lipson, *Freeform Fabrication of Electro active Polymer Actuators and Electromechanical Devices*, Mechanical and Aerospace Engineering, Cornell University, November 2005.
[109] K. Morimoto, A. Utsumi, S. Konishi, A design of longitudinally-divided balloon structure in PDMS pneumatic balloon actuator based on fem simulations, 2011 16th International Solid-State Sensors, Actuators and Microsystems Conference, doi: 10.1109/Transducers.2011.5969597, pp. 2774–2777, August 2011.
[110] S. Leary, Y.B. Cohen, Electrical impedance of ionic polymeric metal composites, Proceedings of SPIE - The International Society for Optical Engineering, doi: 10.1117/12.349713, March 1999.
[111] Y. Shan, K.K. Leang, Application of feed forward dynamics compensation in ionic-polymer metal composite actuators, Proc. SPIE 6927, Electro active Polymer Actuators and Devices (EAPAD), 69270F, doi: 10.1117/12.776659, April 2008.
[112] Y.B. Cohen, Electro-active polymers: current capabilities and challenges, Smart Structures and Materials 2002: Electroactive Polymer Actuators and Devices (EAPAD), Proceedings of SPIE, 4695, 2002.
[113] M. Shahinpoor, M. Mojarrad, Ionic Polymer Sensors and Actuators, United States Patent, Patent No.: US 6,475,639 B2, Date of Patent: Nov. 5, 2002.
[114] H. Sahoo, T. Pavoor, S. Vancheeswaran, Actuators based on electroactive polymers, *Current Science*, 81, 7, pp. 743–746, 2001.
[115] R.K. Jain, S. Datta, S. Majumder, A. Dutta, Two IPMC fingers based micro gripper for handling, *International Journal of Advanced Robotic Systems*, 8, 1, ISSN 1729-8806, pp. 1–9, 2011.
[116] H.P. Monner, Smart materials for active noise and vibration reduction, Keynote Paper, Novem – Noise and Vibration: Emerging Methods Saint-Raphaël, France, 18–21 April, pp. 1–17, 2005.
[117] W.H. Choy, *Study on Fabrication and Performance of IPMCs (Ionic Polymer-Metal Composites)*, B.Tech Thesis, Institute of Textiles & Clothing, The Hong Kong Polytechnic University, 2010.
[118] V. Panwar, C. Lee, S.Y. Ko, J.O. Park, S. Park, Dynamic mechanical, electrical, and actuation properties of ionic polymer metal composites using PVDF/PVP/PSSA blend membranes, *Materials Chemistry and Physics*, 135, pp. 928–937, 2012, doi: 10.1016/j.matchemphys.2012.05.081.
[119] J.D. Carrico, T. Tyler, K.K. Leang, A comprehensive review of select smart polymeric and gel actuators for soft mechatronics and robotics applications: fundamentals, freeform fabrication, and motion control, *International Journal of Smart and Nano Materials*, 8, 4, pp. 144–213, 2017, doi: 10.1080/19475411.2018.1438534.

[120] R. Caponetto, S. Graziani, FulvioL. Pappalardo, F. Sapuppo, Experimental characterization of ionic polymer metal composite as a novel fractional order element, *Advances in Mathematical Physics*, 2013, Article ID 953695, 10 pages, 2013, doi: 10.1155/2013/953695.

[121] S.N. Nasser, C.W. Thomas, Ionomeric polymer-metal composites, *Ionomeric Polymer-Metal Composites*, Mohsen Shahinpoor, (Ed), Royal Society of Chemistry, Chapter 6, pp. 171–230.

[122] B. Bhandari, G.Y. Lee, S.H. Ahn, A review on IPMC material as actuators and sensors: fabrications, characteristics and applications, *International Journal of Precision Engineering and Manufacturing*, 13, 1, pp. 141–163, 2012.

[123] Wikipedia, Ionic Polymer–Metal Composites, https://en.wikipedia.org/wiki/Ionicpolymer_metal_composites, 2003.

[124] L. Yang, D. Zhang, X. Zhang, A. Tian, Fabrication and actuation of cu-ionic polymer metal composite, *Polymers*, 12, 460, 2020, doi: 10.3390/polym12020460.

[125] C.K. Chung, Y.Z. Hong, P.K. Fung, M.S. Ju, C.C.K. Lin, T.C. Wu, A novel fabrication of ionic polymer-metal composites (IPMC) actuator with silver nano-powders, The 13th International Conference on Solid-State Sensors, Actuators and Microsystems, Seoul, Korea, pp. 217–220, June 2005.

[126] C.K. Chunga, P.K. Funga, Y.Z. Honga, M.S. Ju, C.C.K. Lin, T.C. Wu, A novel fabrication of ionic polymer-metal composites (IPMC) actuator with silver nano-powders, *Sensors and Actuators B*, 117, pp. 367–375, 2006, doi: 10.1016/j.snb.2005.11.021.

[127] Z. Chen, D. Hedgepeth, X. Tan, Nonlinear capacitance of ionic polymer-metal composites, *Proceedings of SPIE - The International Society for Optical Engineering*, 7287, 728715-728715-12, March 2009, doi: 10.1117/12.815792.

[128] Z. Chen, Y. Shen, N. Xi, X. Tan, Integrated sensing for ionic polymer–metal composite actuators using PVDF thin films, *Smart Material and Structures*, 16, pp. S262–S271, 2007, doi: 10.1088/0964-1726/16/2/S10.

[129] P. Brunetto, R. Caponetto, G. Dongola, L. Fortuna, S. Graziani, A scalable fractional model for IPMC actuator, PHYSCON 2009, Catania, Italy, September, 2009.

[130] B. Lopes, P.J.C. Branco, Electromechanical characterization of non-uniform charged ionic polymer-metal composites (IPMC) devices, 4th World Congress on Biomimetics, Artificial Muscles and Nano-Bio IOP Publishing Journal of Physics: Conference Series 127–012004, 2008.

[131] K. Tsiakmakis, J. Brufau, M. Puig-Vidal, Th. Laopoulos, Modeling IPMC actuators for model reference motion control, I2MTC 2008 – IEEE International Instrumentation and Measurement Technology Conference Victoria, Vancouver Island, Canada, May 12–15, pp. 1168–1173, 2008.

[132] C. Bonomo, L. Fortuna, P. Giannone, S. Graziani, A sensor-actuator integrated system based on IPMCs, *Proceedings of IEEE*, 1, pp. 489–492, 2004.

[133] M. Porfiri, Charge dynamics in ionic polymer metal composites, *Journal of Applied Physics*, 104, 104915, pp. 104915-1–104915-10, 2008.

[134] K. Park, Evaluation of circuit models for an IPMC (ionic polymer-metal composite) sensor using a parameter estimate method, *Journal of the Korean Physical Society*, 60, 5, pp. 821–829, March 2012.

[135] D. Xue, Z. Chen, L. Hao, X. Xu, Y. Liu, Modeling and control of IPMC for micromanipulation, Proceedings of the 8th World Congress on Intelligent Control and Automation July 6–9 2010, Jinan, China, pp. 2401–2405, 2010.

[136] X. Bao, Y.B. Cohen, S.S. Lih, Measurements and macro models of ionomeric polymer-metal composites (IPMC), *Smart Structures and Materials 2002: Electroactive Polymer Actuators and Devices (EAPAD)*, Y. Bar-Cohen, (Ed), Proceedings of SPIE, 4695, pp. 220–227, 2002.

[137] T. Wallmersperger, A. Horstmann, B. Kroplin, D.J. Leo, Thermodynamical modeling of the electromechanical behavior of ionic polymer metal composites, *Journal of Intelligent Material Systems and Structures*, 20, pp. 741–750, 2009.
[138] N.D. Bhat, *Modelling and Precision Control of Ionic Polymer Metal Composite*. Master's thesis, Texas A&M University.
[139] G. Nishida, M. Sugiura, M. Yamakita, B. Maschke, R. Ikeura, Multi-input multi-output integrated ionic polymer-metal composite for energy controls, *Micromachines*, 3, pp. 126–136, 2012, doi: 10.3390/mi3010126.
[140] C. Jo, D. Pugal, I.K. Oha, K.J. Kim, K. Asaka, Recent advances in ionic polymer–metal composite actuators and their modeling and applications, *Progress in Polymer Science*, 38, 7, pp. 1037–1066, 2013.
[141] Y. Cha, M. Aureli, M. Porfiri, A physics-based model of the electrical impedance of ionic polymer metal composites, *Journal of Applied Physics*, 111, 124901, 2012.
[142] K. Ikeda, M. Sasaki, H. Tamagawa, IPMC bending predicted by the circuit and viscoelastic models considering individual influence of faradaic and non-faradaic currents on the bending, *Sensors and Actuators B*, 190, pp. 954–967, 2014, doi: 10.1016/j.snb.2013.09.016.
[143] D. Bandopadhya, J. Njuguna, Estimation of bending resistance of ionic polymer metal composite (IPMC) actuator following variable parameters pseudo-rigid body model, *Materials Letters*, 63, pp. 745–747, 2009, doi: 10.1016/j.matlet.2008.12.048.
[144] M. Shahinpoor, K.J. Kim, The effect of surface-electrode resistance on the performance of ionic polymer–metal composite (IPMC) artificial muscles, *Smart Materials and Structures*, 9, pp. 543–551, 2000. Printed in the UK.
[145] A. Punning, M. Kruusmaa, A. Aabloo, Surface resistance experiments with IPMC sensors and actuators, *Sensors and Actuators A*, 133, pp. 200–209, 2007.
[146] D. Pugal, K.J. Kim, A. Aabloo, An explicit physics-based model of ionic polymer-metal composite actuators, *Journal of Applied Physics*, 110, 084904, 2011, doi: 10.1063/1.3650903.
[147] K. Kruusamäea, A. Punninga, M. Kruusmaaa, A. Aablooa, Dynamical variation of the impedances of IPMC, Proc. of SPIE, 7287, doi: 10.1117/12.815642.
[148] R. Tiwari, Ionic polymer–metal composite mechanoelectric transduction: effect of impedance, *International Journal of Smart and Nano Materials*, 3, 4, pp. 275–295, 2012, doi: 10.1080/19475411.2012.673511.
[149] Y.B. Cohen, Electroactive polymers as artificial muscles - capabilities, potentials and challenges, *Handbook on Biomimetics*, Y. Osada (Chief Ed.), "Motion" paper #134, NTS Inc., Section 11, in Chapter 8, August 2000.
[150] Y.B. Cohen, X. Bao, S. Sherrit, S.S. Lih, Characterization of the electromechanical properties of ionomeric polymer-metal composite (IPMC), Proceedings of the SPIE Smart Structures and Materials Symposium, EAPAD Conference, San Diego, CA, Paper 4695-33, March 2002.
[151] Z. Chen, Y. Shen, J. Malinak, N. Xi, X. Tan, Hybrid IPMC/PVDF structure for simultaneous actuation and sensing, *Smart Structures and Materials*, 6168, pp. 61681L1–61681L9, 2006.
[152] J. Wpaquette, K.J. Kim, D. Kim, W. Yim, The behavior of ionic polymer–metal composites in a multi-layer configuration, *Smart Materials and Structures*, 14, pp. 881–888, 2005.
[153] B. Kim B.M. Kim, R.H. Oh, Analysis of mechanical characteristics of the ionic polymer metal composites (IPMC) actuator using cast ion-exchange film, Smart Structures and Materials 2003: Electroactive Polymer Actuators and Devices (EAPAD), Proceedings, 5051, pp. 486–495, 2003.

[154] Y. Wu, S.N. Nasser, Verification of micromechanical models of actuation of ionic polymer-metal composites (IPMCs), Smart Structures and Materials 2004: Electroactive Polymer Actuators and Devices (EAPAD), San Diego, CA, USA, Proc. SPIE 5385, 155, 2004.

[155] S.N. Nasser, S. Zamani, Experimental study of Nafion- and Flemion-based ionic polymer metal composites (IPMCs) with ethylene glycol as solvent, Smart Structures and Materials 2003: Electroactive Polymer Actuators and Devices (EAPAD), proceedings of SPIE, 5051, 233, 2003.

[156] Z. Lu, P.C.Y. Chen, W. Lin, Force sensing and control in micromanipulation, *IEEE Transaction on Systems, Man, and Cybernetics – Part C: Applications and Reviews*, 36, 6, pp. 713–724, 2006.

[157] P. Brunetto, L. Fortuna, P. Giannone, S. Graziani, S. Strazzeri, Static and dynamic characterization of the temperature and humidity influence on IPMC actuators, *IEEE Transactions on Instrumentation and Measurement*, 59, pp. 1–16, 2010.

[158] S. Konishi, S. Sawano, S. Kusuda, T. Sakakibara, Fluid-resistive bending sensor compatible with a flexible pneumatic balloon actuator, *Journal of Robotics and Mechatronics*, doi: 10.20965/jrm.2008.p0436, 20, 3, pp. 436–440, June 2008.

[159] J.A. Vickers, *The Development and Implementation of an Ionic Polymer Metal Composite Propelled Vessel Guided by a Goal-Seeking Algorithm*, Texas A&M University, Master of Science, May 2007.

[160] S.N. Nassera, S. Zamani, Y. Tor, effect of solvents on the chemical and physical properties of ionic polymer-metal composites, *Journal of Applied Physics*, 99, 104902, pp. 104902-1–104902-17, May 2006, doi: 10.1063/1.2194127.

[161] Z. Chen, K.Y. Kwon, X. Tan, Integrated IPMC/PVDF sensory actuator and its validation in feedback control, *Sensors and Actuators*, 144, 2, pp. 231–241, 2008.

[162] X. Liu, K. Kim, Y. Zhang, Y. Sun, Nano newton force sensing and control in microrobotic cell manipulation, *The International Journal of Robotics Research*, 28, 8, pp. 1065–1076, 2009.

[163] K. Onishi, S. Sewa, K. Asaka, N. Fujiwara, K. Oguro, Morphology of electrodes and bending response of the polymer electrolyte actuator, *Electrochimica Acta*, 46, pp. 737–743, 2000.

[164] C.S. Kothera, *Micro-Manipulation and Bandwidth Characterization of Ionic Polymer Actuators*, Virginia Polytechnic Institute and State University, Blacksburg, Virginia, Master of Science, December 2002.

[165] R.K. Jain, A. Datta, S. Majumder, Design and control of an IPMC artificial muscle finger for micro gripper using EMG signal, *International Journal of Mechatronics*, 23, pp. 381–394, 2013.

[166] J. Sunghee, S. Jeomsik, K. Gyuseok, L. Sukmin, M. Museong, Effects of the electrode interface on the electric properties of IPMC for artificial muscles, IFMBE Proceedings Vol. 14/5, 5, Track 16, pp. 2973–2976, 2007.

[167] B. Ansafa, T.H. Duonga, N.I. Jaksica, J.L. DePalmaa, A.H.A. Allaqa, M.B.M. Deherreraa, B. Lia, Influence of humidity and actuation time on electromechanical characteristics of ionic polymer-metal composite actuators, Procedia Manufacturing, 28th International Conference on Flexible Automation and Intelligent Manufacturing (FAIM2018), 17, pp. 960–967, June 2018.

[168] D. Pugala, A. Aabloo, K.J. Kim, Dynamic surface resistance model of IPMC, Proc. of SPIE, 7289, doi: 10.1117/12.815824, pp. 72891E-1–72891E-9, 2009.

[169] W.H. Mohdlsa, A. Hunt, S. Hassan HosseinNia, Sensing and self-sensing actuation methods for ionic polymer–metal composite (IPMC): a review, *Sensors*, 19, 3967, pp. 1–36, 2019, doi: 10.3390/s19183967, www.mdpi.com/journal/sensors.

[170] M. Amirkhani, P. Bakhtiarpour, *Ionic Polymer Metal Composites: Recent Advances in Self-Sensing Methods*, Published on 19 2015 on https://pubs.rsc.org, doi: 10.1039/-9781782627234-00240, Chapter 19, pp. 240–256, 2016.
[171] A. Punning, M. Kruusmaa, A. Aabloo, A self-sensing ion conducting polymer metal composite (IPMC) actuator, *Sensors and Actuators A*, 136, pp. 656–664, 2007, doi: 10.1016/j.sna.2006.12.008.
[172] B. Koo, D.S. Na, S. Lee, Control of IPMC actuator using self-sensing method, *IFAC Proceedings*, 42, 3, pp. 267–270, 2009, doi: 10.3182/20090520-3-KR-3006.00043.
[173] B. Ko, H. Kwon, S. Lee, A self-sensing method for IPMC actuator, *Advances in Science and Technology*, 56, pp. 111–115, 2008, doi: 10.4028/www.scientific.net/AST.56.111.
[174] M. Sasaki, W. Lin, H. Tamagawa, S. Ito, K. Kikuchi, Self-sensing control of Nafion-based ionic polymer-metal composite (IPMC) actuator in the extremely low humidity environment, *Actuators*, 2, pp. 74–85, 2013, doi: 10.3390/act2040074.
[175] Z. Chen, T.I. Um, H.B. Smith, A novel fabrication of ionic polymer–metal composite membrane actuator capable of 3-dimensional kinematic motions, *Sensors and Actuators A: Physical*, 168, 1, pp. 131–139, July 2011, doi: 10.1016/j.sna.2011.02.034.
[176] M. Yu, Q.S. He, Y. Ding, D. Guo, J.B. Li, Z.D. Dai, Force optimization of ionic polymer metal composite actuators by an orthogonal array method, *Chinese Science Bulletin*, 56, 19, pp. 2061–2070, July 2011, doi: 10.1007/s11434-011-4509-9.
[177] S.G. Lee, H.C. Park, S.D. Pandita, Y. Yoo, Performance improvement of IPMC (ionic polymer metal composites) for a flapping actuator, *International Journal of Control, Automation, and Systems*, 4, 6, pp. 748–755, 2006.
[178] K. Jung, J. Namb, H. Choi, Investigations on actuation characteristics of artificial muscle actuator, *Sensors and Actuators*, 107, pp. 183–192, 2003.
[179] G.H. Fen, R.H. Chen, Fabrication and characterization of arbitrary shaped µIPMC transducers for accurately controlled biomedical applications, *Sensors and Actuators*, 143, pp. 34–40, 2008.
[180] K. Tsiakmakis, J.B. Penella, M.P. Vidal, T. Laopoulos, A camera based method for the measurement of motion parameters of IPMC actuators, *IEEE Transactions on Instrumentation and Measurement*, 58, 8, pp. 2626–2633, August 2009.
[181] K. Sadeghipour, R. Salomon, S. Neogi, Development of a novel electrochemically active membrane and 'smart' material based vibration sensor/damper, *Smart Material Structure*, 1, 2, pp. 172–179, 1992.
[182] J.H. Lee, J.S. Oh, G.H. Jeong, J.Y. Lee, B.R. Yoon, J.Y. Jho, K. Rhee, New computational model for predicting the mechanical behavior of ionic polymer metal composite (IPMC) actuators, *International Journal of Precision Engineering and Manufacturing*, 12, 4, pp. 737–740, 2011.
[183] Q. Chen, K. Xiong, K. Bian, N. Jin, B. Wang, Preparation and performance of soft actuator based on IPMC with silver electrodes, *Frontiers of Mechanical Engineering in China*, 4, 4, pp. 436–440, 2009.
[184] Q. He, M. Yu, L. Song, H. Ding, X. Zhang, Z. Dai, Experimental study and model analysis of the performance of IPMC membranes with various thickness, *Journal of Bionic Engineering*, 8, 1, pp. 77–85, 2011.
[185] X. Chen, G. Zhu, X. Yang, D.L.S. Hung, X. Tan, Model-based estimation of flow characteristics using an ionic polymer–metal composite beam, *IEEE/ASME Transaction on Mechatronics*, 18, 3, pp. 932–943, 2013.
[186] R.C. Richardson, M.C. Levesley, M.D. Brown, J.A. Hawkes, K. Watterson, P.G. Walker, Control of ionic polymer metal composites, *IEEE/ASME Transactions on Mechatronics*, 8, 2, pp. 245–253, 2003.

[187] K. Onishi, S. Sewa, K. Asaka, N. Fujiwara, K. Oguro, The effects of counter ions on characterization and performance of a solid polymer electrolyte actuator, *Electrochimica Acta*, 46, pp. 1233–1241, 2001.

[188] G. Alici, N.N. Huynh, Performance quantification of conducting polymer actuators for real applications: a microgripping system, *IEEE/ASME Transactions on Mechatronics*, 12, 1, pp. 73–84, 2007.

[189] R.K Jain, U.S Patkar, S. Majumdar, Micro gripper for micromanipulation using IPMCs (ionic polymer metal composites), *Journal of Scientific & Industrial Research*, 68, pp. 23–28, 2009.

[190] R. Lumia, M. Shahinpoor, IPMC microgripper research and development, 4th World Congress on Biomimetics, Artificial Muscles and Nano-Bio IOP Publishing Journal of Physics: Conference Series, 127, 012002, 2008.

[191] E. Biddiss, T. Chaua, Electroactive polymeric sensors in hand prostheses: bending response of an ionic polymer metal composite, *Medical Engineering & Physics*, 28, pp. 568–578, 2006.

[192] K. Takagi, N. Kamamichi, B. Stoimenov, K. Asaka, T. Mukai, Z.W. Luo, Frequency response characteristics of IPMC sensors with current/voltage measurements, Proc. of SPIE, 6927, doi: 10.1117/12.776189, pp. 692724-1 to 692724-10, 2008.

[193] V. Panwar, K. Cha, J.O. Park, S. Park, High actuation response of PVDF/PVP/PSSA based ionic polymer metal composites actuator, *Sensors and Actuators B*, 161, pp. 460–470, 2012, doi: 10.1016/j.snb.2011.10.062.

[194] L. Yang, D. Zhang, X. Zhang, H. Wang, Property of Nafion-ionic polymer-metal composites based on Mori–Tanaka methodology and gradient mechanics, *Applied Physics A*, 126, 633, July 2020, doi: 10.1007/s00339-020-03807-9.

[195] J.D. Carrico, T. Hermans, K.J. Kim, K.K. Leang, 3D-printing and machine learning control of soft ionic polymer-metal composite actuators, *Scientific Reports*, 9, 17482, 2019, doi: 10.1038/s41598-019-53570-y.

[196] V. De Luca, P. Digiamberardino, G.D. Pasquale, S. Graziani, A. Pollicino, E. Umana, M.G. Xibilia, Ionic electroactive polymer metal composites: fabricating, modelling and applications of post-silicon smart devices, *Journal of Polymer Science Part B Polymer Physics*, 51, pp. 699–734, May 2013, doi: 10.1002/polb.23255.

[197] P. Bakhtiarpour, A. Parvizi, M. Müller, M. Shahinpoor, O. Marti, M. Amirkhani, An external disturbance sensor for ionic polymer metal composite actuators, *Smart Materials and Structures*, 25, 015008, 7 pp., 2016, doi: 10.1088/0964-1726/25/1/015008.

[198] S.N. Nasser, Y. Wu, Tailoring actuation of ionic polymer-metal composites through cation combination, Proceedings Volume 5051, Smart Structures and Materials 2003: Electroactive Polymer Actuators and Devices (EAPAD), 2003, doi: 10.1117/12.484439.

[199] J.L. Gonzalez, A. Rubio, Francesc Moll, Human powered piezoelectric batteries to supply power to wearable electronic devices, *International Journal of the Society of Materials Engineering for Resources*, 10, 1, pp. 34–40, March 2002.

[200] M. Meddad, A. Eddiai, A. Chérif, A. Hajjaji, Y. Boughaleb, Model of piezoelectric self-powered supply for wearable devices, *Superlattices and Microstructures*, 71, pp. 105–116, 2014. www.elsevier.com/locate/superlattices, doi: 10.1016/j.spmi.2014.03.038.

[201] Y.B. Cohen, Electroactive polymers (EAP) as artificial muscles, JPL, 818-354-2610, yosi@jpl.nasa.gov, http://ndeaa.jpl.nasa.gov/, Karman Auditorium Lecture Series, February 2002.

[202] M. Shahinpoor, Y. Bar-Cohen, T. Xue, J.O. Simpson, J. Smith, Ionic polymer-metal composites (IPMC) as biomimetic sensors and actuators-artificial muscles, Proceedings of SPIE's 5Ih Annual International Symposium on Smart Structures and Materials, 1–5 March, 1998, San Diego, CA. Paper No. 3324-27, 1998.

[203] M Shahinpoor, Y Bar-Cohen, JO Simpson, J Smith, Ionic polymer–metal composites (IPMCs) as biomimetic sensors, actuators and artificial muscles—a review, *Smart Materials and Structures*, 7, pp. R15–R30, 1998.

[204] Y. Ming, Y. Yang, R. P. Fu, C. Lu, L. Zhao, Y. M. Hu, C. Li, Y. X. Wu, H. Liu and W. Chen, "IPMC sensor integrated smart glove for pulse diagnosis, braille recognition, and human-computer interaction", Advanced Materials Technologies, 3, pp. 1800257 (1–8), 2018, doi: 10.1002/admt. 201800257.

[205] B.H. Kang, J.T. Wen, Design of compliant mems grippers for micro-assembly tasks, Proceedings of the 2006 IEEE/RSJ International Conference on Intelligent Robots and Systems October 9–15, 2006, Beijing, China, pp. 760–765, 2006.

[206] L.M. Enache, I. Bunescu, Air ionization - an environmental factor with therapeutic potential, *GEO Review*, 29, pp. 31–39, 2019.

[207] O.A. Albrechtsen, V. Clausen, F.G. Christensen, J.G. Jensen, T. Mealier, The influence of small atmospheric ions on human well-being and mental performance, *International Journal of Biometeorology*, 22, 4, pp. 249–262, 1978.

[208] P. Brunetto, L. Fortuna, P. Giannone, S. Graziani, S. Strazzeri, Static and dynamic characterization of the temperature and humidity influence on IPMC actuators, *IEEE Transactions on Instrumentation and Measurement*, 59, pp. 893–908, 2009.

[209] J.B. Penella, M.P. Vidal, P Giannone, S Graziani, S Strazzeri, Characterization of the harvesting capabilities of an ionic polymer metal composite device, *Smart Materials and Structures*, 17, 015009, 15pp, 2008, doi:10.1088/0964-1726/17/01/015009.

[210] S. Bhattacharya, B. Bepari, S. Bhaumik, IPMC-actuated compliant mechanism-based multifunctional multifinger microgripper, *Mechanics Based Design of Structures and Machines*, 42, pp. 312–325, 2014, ISSN: 1539-7734 print/1539-7742 online, doi: 10.1080/15397734.2014.899912.

8 Future Directions on IPMC Research

Srijan Bhattacharya
RCC Institute of Information Technology

CONTENTS

8.1 Introduction 193
8.2 Future Research Objectives and Methodology 194
    8.2.1 Study of the Characteristics of IPMC 194
    8.2.2 IPMC Sensor Parameters Study 195
    8.2.3 Mechatronics Applications of IPMC with Different Shape of Object Grasp Capability by Human Finger 195
    8.2.4 Design and Development of IPMC Analogous to EMG Sensors for Biomedical Applications 195
    8.2.5 Design and Development of IPMC-Based Taste Sensors 196
    8.2.6 Bioinstrumentation Applications of IPMC for Moisture Capturing Capability in Space 196
    8.2.7 3D Printing of IPMC 196
8.3 Summary 196
References 197

8.1 INTRODUCTION

A high bending force output at low actuation voltage (1–5 V DC) is the significant electromechanical property of ionic polymer–metal composites (IPMCs), a special category of EAPs. Ion exchange/diffusion is the driving force for this actuation, which creates an electric field across the polymer membrane. The processing speed is very fast (milliseconds) for this type of ion exchange EAPs [1,2]. Electroplating with platinum (Pt) on both sides of the perfluorinated polymer membrane having ion exchange property is accomplished for metal lead electroding [2]. Applications of soft robotic gripping in both macro- and micro-size object grasping are significant features for micromanipulation based on IPMC artificial finger actuators [3]. The stable gripping capability of the IPMC actuated end effector for microgripping is within the range of 1–10 mm diameter [4]. The modeling of IPMC strip bending pattern using a 20-link hyper-redundant serial manipulator was developed by Chattaraj et al. [5]. Inverse kinematics model based on Tractrix was introduced with the bending profile while distilled water used with 1.5 N LiCl and NaCl solutions, respectively. Cantilever deflection theory was substituted by an alternative approach

DOI: 10.1201/9781003204664-8

for the modeling and detection of IPMC bending profile proposed by Chattaraj et al. [6]. Inverse kinematics algorithm using Tractrix-based hyper-redundant approach is adopted when the IPMC actuator is soaked with LiCl and distilled water as solvents. The conventional two-jaw model along with modified passive jaw gripping actuators were studied in the above media. The phenomenon of voltage change under the human skin due to the spontaneous muscular movement implies the electrochemical properties of human muscles. Electromyogram (EMG) sensors are electrical-type sensors, developed by the principle of detection of the change in voltage due to muscular activities into EMG signal generation [7]. The movement of muscles and nerve cells provides the diagnostic feature of EMG signals for human health condition monitoring. The change in the generated voltage due to muscular movement which the IPMC strips can detect by the difference in pressure generated on the membrane surface follows the analogy of IPMC electrodes with EMG electrodes [8]. So, IPMC-compatible biosensors are potentially significant to retrieve the diagnostic information of muscular signals for the pre-prediction of disease by analyzing the EMG signals due to muscular motion. Machine learning models for data training and feature extraction in this regard will be adopted for optimum accuracy in view of data analysis Das et al. [9]. The sensing characteristics of the IPMC strips are applied to sense five general tastes (sweet, sour, bitter, salty and spicy) along with security and safety of food content. IPMCs are immersed in individual solutions for certain duration for doping action. Thus, different solutions having distinct concentration of ions will lead to unique identifications by creating distinct bending or deformation profile of the IPMC strip. Different analogous voltages in the range of millivolts are generated across the membranes for different solutions De et al. [10]. Finally, 3D-printed IPMCs are published and they can mimic octopus, caterpillar, etc. [11]. This research opens a new horizon of IPMC applications.

8.2 FUTURE RESEARCH OBJECTIVES AND METHODOLOGY

- Study of the characteristics of different shapes and sizes of IPMC as actuators and sensors.
- Design and development of IPMC analogous to EMG sensors for biomedical applications.
- Mechatronics applications of IPMC with different object/shape identification abilities.
- Design and development of IPMC-based taste sensors.
- Bioinstrumentation applications of IPMC for moisture capturing in space.
- 3D manufacturing of IPMC.

8.2.1 STUDY OF THE CHARACTERISTICS OF IPMC

- For this, the IPMC can be actuated by giving the input signal of 1–5 V and 1–15 Hz frequency, through indigenous power supply circuit. Otherwise, SourceMeter KEITHLEY 2400-LV (20 V, 1 A, 20 W)/other power supplies can be used for the power supply to IPMC.

Future Directions on IPMC Research

- To measure the blocking force exerted by the IPMC, a load cell will be used (Model GM2, PTC Electronics Inc., Wyckoff, NJ, the USA) because it can measure 300 mN force, pan diameter 10 cm.
- And to measure the displacement, a laser displacement sensor KEYENCE LB-12 will be used.

8.2.2 IPMC Sensor Parameters Study

- IPMC samples of square or rectangular shape with dimensions 30 mm×30 mm or 50 mm×30 mm, or a 20 mm/30 mm diameter circular IPMC can be taken (or any other sample as per the objective).
- Those samples will be clipped, and each sample will be treated as a 3×3 matrix and a 5×3 matrix (or any other sample as per the objective).
- Force will be applied to each element of the matrix to find out the potential change sensed by those IPMCs during the application of force in different positions on the IPMC.
- Electroding of IPMC for each element will be prepared with an insulated, coated copper/aluminum wire.
- The above-mentioned process will be applied for uncovered IPMCs.
- For encapsulating the IPMC, it will be shielded with the silicon rubber polydimethylsiloxane (PDMS).
- Electroding of IPMC for this arrangement will be done before PDMS is encapsulated.

8.2.3 Mechatronics Applications of IPMC with Different Shape of Object Grasp Capability by Human Finger

- An IPMC sample of 30 mm×30 mm or 50 mm×30 mm square or rectangular shape, or a 20-mm-/30-mm-diameter circular IPMC will be placed on human forearm, and the response will be recorded on a digital storage oscilloscope (DSO). These results will be stored (or any other sample as per the objective).
- Using MATLAB and electronic circuit (using microcontroller/NI data card), the same signal will be recorded in real time.
- After this experimental arrangement, the signals of different shapes of objects/activities (for example, square, circular, cylindrical, writing with pen, applying pressure on table/wall) will be recorded.

8.2.4 Design and Development of IPMC Analogous to EMG Sensors for Biomedical Applications

- The dimension of the IPMC strip will be 45×15×0.35 mm—Nafion membrane electroplated with noble metal Pt/Au/Ag.
- The IPMC output will be fed to an AD 620 high-precision, high-gain and low-power instrumentation amplifier for noise elimination and IPMC voltage amplification.

- The standard EMG electrode will be used, and its output will be fed to Texas Instruments' TL 082 JFET having a wide bandwidth and low input bias current (50 pA) for EMG signal amplification.
- The IPMC and EMG electrodes will be placed on the bicep muscles for differential input voltages, and the reference electrode will be placed on the elbow.
- PCI-1711 data acquisition card and LabVIEW will be used for data logging. Statistical analysis and time domain feature extraction will be done by Python programming using machine learning tools for the similarity analysis between IPMC and EMG.

8.2.5 Design and Development of IPMC-Based Taste Sensors

- A rectangular-shaped IPMC strip of $27 \times 18 \times 0.2$ mm dimension will be immersed in different solutions containing food extracts of different tastes ((1) salty, (2) sour, (3) spicy, (4) bitter and (5) sweet). Salt, lemon, chili, bitter gourd and sugar, respectively, were used for making the solutions.
- Different degrees of bending of the IPMC strip w.r.t the null position (both clockwise and counterclockwise) will be monitored by the precision laser displacement sensor KEYENCE LB-12.
- Different ranges of voltage generation (mV) along with the strip movement will be monitored by a digital voltmeter.

8.2.6 Bioinstrumentation Applications of IPMC for Moisture Capturing Capability in Space

An IPMC specimen of dimensions $45 \times 15 \times 0.35$ mm—Nafion membrane electroplated with noble metal Pt/Au/Ag—will be tested to find the moisture capturing capacity in space or an equivalent environment.

8.2.7 3D Printing of IPMC

- Methods of 3D printing of IPMCs. First, the fused filament fabrication process requires manufacturing a filament using Nafion (or Aquivion-49).
- Next, a custom-designed 3D printer is required to process the precursor material for 3D printing.
- Filament extrusion and 3D printing of ionomeric precursor.
- Functionalization process. After printing, the component has to be made electroactive through a functionalization process.
- Electrode plating process.

8.3 SUMMARY

IPMC technology-based sensors and transducers are considered for the research work to develop MEMS in different environmental conditions. Object identification and microgripping by implementing soft robotics, EMG signal interfacing and

Future Directions on IPMC Research

analysis using IPMC membranes, taste sensing application and moisture capturing capabilities of IPMCs at different atmospheric conditions can be taken under investigation in this current research work. This chapter intends to act as an extensive resource in the field of IPMC sensor and actuator technology and their applications.

It is suggested that the readers follow Chapter 1 also for a more detailed idea of IPMCs and their applications.

REFERENCES

[1] S. Bhattacharya, B. Bepari, and S. Bhaumik, IPMC-Actuated Compliant Mechanism-Based Multifunctional Multifinger Microgripper, *Mechanics Based Design of Structures and Machines*, Vol. 42, No. 3, pp. 312–325, 2014, ISSN: 1539-7734 print/1539-7742 online, DOI: 10.1080/15397734.2014.899912.

[2] M. Shahinpoor and K. J. Kim, Ionic Polymer-Metal Composites: I. Fundamentals, *Smart Materials and Structures*, Vol. 10, No. 4, pp. 819–833, August 2001, DOI: 10.1088/0964-726/10/4/327.

[3] S. Bhattacharya, S. Khan, T. Sil, B. Bepari and S. Bhaumik, IPMC Based Data Glove for Finger Motion Capturing, 2nd Int. Conf. of Robotics Society of India, BITS Goa, India, July 2–4th, 2015.

[4] S. Bhattacharya, R. Chattaraj, M. Das, A. Patra, B. Bepari and S. Bhaumik, Simultaneous Parametric Optimization of IPMC Actuator for Compliant Gripper, *International Journal of Precision Engineering and Manufacturing*, Vol. 16, No. 11, pp. 2289–2297, 2015, DOI: 10.1007/s12541-015-0294-8.

[5] R. Chattaraj, S. Khan, S. Bhattacharya, B. Bepari, D. Chatterjee and S. Bhaumik, Development of Two Jaw Compliant Gripper Based on Hyper-Redundant Approximation of IPMC Actuators, *Sensors and Actuators A: Physical*, Vol. 251, pp. 207–218, 2016, ISSN: 0924-4247, DOI: 10.1016/j.sna.2016.10.017.

[6] R. Chattaraj, S. Khan, S. Bhattacharya, B. Bepari, D. Chatterjee and S. Bhaumik, Shape Estimation of IPMC Actuators in Ionic Solutions Using Hyper Redundant Kinematic Modeling, *Mechanism and Machine Theory*, Vol. 103, pp. 174–188, September 2016, DOI: 10.1016/j.mechmachtheory.2016.05.002.

[7] A. Paul, N. Dey and S. Bhattacharya, Similarity Analysis of IPMC and EMG Signal with Comparative Study of Statistical Features, *Advancements in Instrumentation and Control in Applied System Applications*, Vol. 1, pp. 1–16, March 2020, DOI: 10.4018/978-1-7998-2584-5.ch001.

[8] S. Bhattacharya, S. Halder, A. Sadhu, S. Banerjee, S. Sinha, S. Banerjee, S. Kundu, B. Bepari and S. Bhaumik, Characteristics of Ionic Polymer Metal Composite (IPMC) as EMG Sensor, *Advancements in Instrumentation and Control in Applied System Applications*, pp. 98–107, March 2020, DOI: 10.4018/978-1-7998-2584-5.ch006.

[9] S. Das, S. Ghosh, R. Guin, A. Das, B. Das, S. Saha, S. Bhattacharya, B. Bepari and S. Bhaumik, IPMC as EMG Sensor to Diagnose Human Arm Activity, 1st International Conference on Industrial Instrumentation & Control, 20–22 August 2021. (Accepted).

[10] A. De, A. Pal, D. Ash, K. Mondal, K. Das, P. Dhar, R. Chakraborty, P. Rakshit, S. Bhattacharya, B. Bepari and S. Bhaumik, Taste Sensor Using Ionic Polymer Metal Composite, In *IEEE Sensors Letters*, Vol. 5, No. 4, pp. 1–4, Article number: 5500304, April 2021, DOI: 10.1109/LSENS.2021.3061546.

[11] C. James, D. H. Tucker, J. K. Kwang and K. L. Kam, 3D-Printing and Machine Learning Control of Soft Ionic Polymer-Metal Composite Actuators, *Scientific Reports*, Vol. 9, pp. 1–17, Article number: 17482, 2019.

Index

3D printing 196

actuation mechanism (IPMC) 18
Ag-IPMC 17, 23
    fabrication Ag-IPMC 25, 26, 27
air–earth current 156, 158

Bepari, Bikash 95, 123
Bhattacharya, Srijan 1, 95, 123, 149, 193
Bhaumik, Subhasis 95, 123
Biswal, Dillip Kumar 17

characterization 5, 25
    Ag-IPMC 63–64, 72, 74, 76
Chattaraj, Ritwik 95
Chattopadhyay, Subrata 149
chemical decomposition method 20
compliant microgripper 9
conductivity 64, 156
control IPMC 7

Das, Suman 149
Duy, Vinh Nguyen 31

ELECTRE II 133
electrical characterization IPMC 5
electric field 158
electroactive polymers (EAPs) 2, 17, 32, 38, 117
electrochemical characterization 72
electromechanical characterization 74
EMG 82, 196
entropy 140
ethylene–propylene diene monomer (EPDM) 138
ethylene–propylene monomer (EPM) 138
ethylene–propylene terpolymer (EPT) 139
ethylene–vinyl acetate (EVA) 138

fabrication (IPMC) 20, 23
    multi-layered Ag-IPMC 38, 60, 166, 168
    silver nano powders 168
    single layered Ag-IPMC 24
force 2, 5, 8, 36, 78, 168, 171, 195
Fourier transform infrared (FTIR) 66
frequency 4, 5, 44, 168, 172, 194

Gare, Gautam Rajendrakumar 95
global electric circuit 153

gripper 97, 113, 114, 125

Ho, Nang Xuan 31
hyper-redundant kinematic modelling 113

IEC (ion exchange capability) 65
inverse kinematics 102, 103
ionic conductivity 64
ionic polymer–metal composite (IPMC) 1, 32, 152
IPMC actuation mechanism 18
IPMC gripper 8, 11, 113
IPMC in space 10, 196

Jacobian transpose 102
Jain, Ravi Kant 53

Khan, Siladitya 95
Kim, Hyung-Man 31

Maxwell current 156
mechanical plating method 21
micro assembly 78, 85
micro gripper 8, 78, 81, 113, 125
microstructure analysis 25
morphological analysis 25
multi-criteria decision making (MCDM) 128
multi-objective optimization 132

ocean environmental conditions 39
ocean kinetic energy-harvesting modules 49
ocean wave kinetic energy 35

PC (conductivity) 65
polydimethylsiloxane (PDMS) 1, 9, 113, 125, 139, 171, 195
polyurethane (PU) 139
polyvinylidene fluoride (PDVF) 139
pseudo-inverse solution 102

remote center compliance (RCC) assembly 78

self-sensing actuation (SSA) 170
space (IPMC) 10, 196

taste sensors 196
tensile strength 67
Tractrix 103

Tractrix based inverse kinematics 103
two-finger microgripper 81

water uptake (WU) 65
wave direction and frequency 44

wearable electronic device 173
wetted surface area and mass 48

Young's modulus 5, 67, 168, 172